# Acknowledgments

The authors and editor wish to thank the following individuals who have contributed to the development of this *Camp and Club Science Sourcebook: Activities and Planning Guide for Science Outside School.*

## Terrific Science Press Design and Production Team

*Document Production Managers:* Susan Gertz, Amy Stander
*Production Coordinator:* Dot Lyon
*Technical Writing:* Dot Lyon, Tom Schaffner, Don Robertson
*Technical Editing:* Dot Lyon, Amy Stander
*Production:* Anita Winkler, Dot Lyon, Tom Schaffner, Don Robertson, Jeri Moore
*Illustrations:* Carole Katz, Tom Schaffner, Anita Winkler
*Cover Design, Layout, and Photography:* Susan Gertz
*Deputy Project Director:* Marina Canepa

## Reviewers

James Arbogast, GOJO Industries, Akron, OH
Cara Bondi, GOJO Industries, Akron, OH
Rachel Cherico, GOJO Industries, Akron, OH
Kathleen Dixon, Department of Molecular and Cellular Biology, The University of Arizona, Tucson, AZ
Mary Beth Genter, Department of Environmental Health, University of Cincinnati, Cincinnati, OH
Jeffrey Johnson, Department of Environmental Health and Safety, Miami University, Oxford, OH
Edward P. Knepp, Department of Microbiology, Miami University, Middletown, OH
Stephen Kralovic, Department of Internal Medicine, University of Cincinnati, Cincinnati, OH
Gretchen A. Mayer, Kettering Medical Center, Kettering, OH
Matt Nance, Department of Chemistry and Biochemistry, Miami University, Oxford, OH
Elisabeth Portman, Kao Brands Company, Cincinnati, OH
Jerry Sarquis, Department of Chemistry and Biochemistry, Miami University, Oxford, OH
Keith Zook, Procter & Gamble, Cincinnati, OH

# Contents

**Getting the Most from This Guide** .................................................................................... vi
**Key Features** ........................................................................................................................ vi

**Planning Your Event** ............................................................................................................ 1
    Planning Made Easy ........................................................................................................ 3
    Example Materials ......................................................................................................... 17

**Topic 1: Disease Control Is in Your Hands**   *camper notebook* .....33   *leader guide* .... 135
    Why Soap? ................................................................................ 34 ............................136
    Design Your Own Soap ........................................................... 37 ............................140
    DNA from Strawberries .......................................................... 39 ............................143
    Colorful Lather Printing ......................................................... 42 ............................147
    Surfactants ............................................................................... 44 ............................150
    Blowing Bubbles ..................................................................... 47 ............................154
    Take-Home Activity: Watching Granny Smith Rot ............. 49 ............................158
    Take-Home Activity: How Clean Is Your Clan? .................. 52 ............................159

**Topic 2: Water Purification**   *camper notebook* .....55   *leader guide* .... 161
    Back to Its Elements ............................................................... 56 ............................162
    Distinguishing Water Samples .............................................. 58 ............................166
    Osmosis with Eggs ................................................................. 61 ............................170
    Water Taste Test ..................................................................... 65 ............................175
    Investigating Mineral Content .............................................. 67 ............................179
    How Hard Is Your Water? ..................................................... 69 ............................182
    Purifying Surface Water ........................................................ 71 ............................185
    How Much Iodine Is Present? ............................................... 74 ............................189
    Removal of Iodine from Water .............................................. 76 ............................193
    Take-Home Activity: What's Really in the Bottle? .............. 78 ............................197
    Take-Home Activity: The Amazing Water Maze ................ 84 ............................198

**Topic 3: Healthy Air**   *camper notebook* .....85   *leader guide* .... 199
    Look at One Part per Million ................................................ 86 ............................200
    Balloon Challenge .................................................................. 88 ............................203
    Search for One Part per Million ............................................ 90 ............................206
    Humidity Detector ................................................................. 91 ............................209
    How Much Oxygen Is in Air? ................................................ 93 ............................212
    Pour a Gas ............................................................................... 96 ............................216
    Pour More Gas ........................................................................ 98 ............................219

# Camp and Club Science Sourcebook:
## Activities and Planning Guide for Science Outside School

### Contributing Authors

Susan Hershberger, Center for Chemistry Education, Miami University
Lynn Hogue, Center for Chemistry Education, Miami University

### Series Editor

Mickey Sarquis, Director
Center for Chemistry Education, Miami University

Terrific Science Press
Miami University Middletown
Middletown, Ohio USA

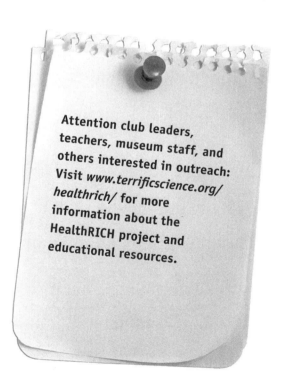

Attention club leaders, teachers, museum staff, and others interested in outreach: Visit www.terrificscience.org/healthrich/ for more information about the HealthRICH project and educational resources.

Terrific Science Press
Miami University Middletown
4200 East University Boulevard
Middletown, OH 45042
513/727-3269
cce@muohio.edu
www.terrificscience.org

© 2008 by Terrific Science Press™
All rights reserved
Printed in the United States of America

10  9  8  7  6  5  4  3  2  1

This monograph is intended for use by teachers, club leaders, museum staff, and properly supervised children and their families. The safety reminders associated with experiments and activities in this publication have been compiled from sources believed to be reliable and to represent the best opinions on the subject as of the date of publication. No warranty, guarantee, or representation is made by the authors or by Terrific Science Press as to the correctness or sufficiency of any information herein. Neither the authors nor the publisher assume any responsibility or liability for the use of the information herein, nor can it be assumed that all necessary warnings and precautionary measures are contained in this publication. Other or additional information or measures may be required or desirable because of particular or exceptional conditions or circumstances.

ISBN 978-1-883822-48-4

The publisher takes no responsibility for the use of any materials or methods described in this monograph, nor for the products thereof. Permission is granted to copy the materials for use by an organization.

This publication was made possible by Grant Number 1 R25 RR16301-01A1 from the National Center for Research Resources (NCRR), a component of the National Institutes of Health (NIH). Its contents are solely the responsibility of the authors and do not necessarily represent the official views of NCRR or NIH.

Department of Health and Human Services
National Institutes of Health

Supported by a Science Education Partnership Award (SEPA)
from the National Center for Research Resources

www.terrificscience.org/healthrich/

    Stirring Indoor Air .................................................................. 100 ..................................222
    Trapping Particulates ............................................................ 102 ..................................225
    Cartesian Diver ....................................................................... 104 ..................................228
    Take-Home Activity: Growing Mold ................................... 106 ..................................231

**Topic 4: Healthy Skin**                       *camper notebook* .... **109**    *leader guide* .... **233**
    Hydrophobic Art .................................................................... 110 ..................................234
    Visible Light Challenge ........................................................ 112 ..................................237
    How Sensitive Is Your Skin? ................................................ 113 ..................................239
    Make and Test Lip Balm ...................................................... 115 ..................................242
    Cover Up, Screen, or Block? ............................................... 119 ..................................247
    Sunning Straws ..................................................................... 122 ..................................252
    Suntan in a Bottle ................................................................. 124 ..................................255
    A Look at Bleaching ............................................................. 127 ..................................259
    Take-Home Activity: UV Detective Challenge ................. 130 ..................................263

**References** .............................................................................................................................. **265**

# Getting the Most from This Guide

This how-to guide provides all the information you need to organize and conduct a successful informal science experience such as a summer camp, weekend science retreat, after-school science program, or museum experience. This program is well-suited for middle school students but is easily modified for children of different ages.

Through fun, engaging hands-on discovery in four main topics, kids will explore some of the personal choices that are part of their everyday lives:

- **Topic 1: Disease Control Is in Your Hands**—healthy disease-control choices related to soap, hand washing, and bacterial protection

- **Topic 2: Water Purification**—healthy drinking water choices such as bottled versus tap water and unfiltered versus filtered water

- **Topic 3: Healthy Air**—healthy indoor air choices related to building ventilation, indoor humidity, and air filtering

- **Topic 4: Healthy Skin**—healthy skin choices related to UV protection, skin bleaching, and skin tanning

> For consistency throughout this document, the informal science experience is called a camp. However, this guide can be used for any type of informal event.

# Key Features

**Planning Your Event: The first section of the guide provides resources to help in planning your event.**

- The activities in this guide will fill at least five half-day sessions (one week-long camp of approximately three-hour days). The planning section provides a sample one-week schedule. For a single-session event or weekend format, use fewer activities in shorter sessions and/or limit the topics.

- Decide on a catchy name for your camp that will appeal to the kids you are targeting. Ideas include *Awesome Science!; So Much Science! So Many Choices!; Science that Gets In Your Face;* and *Science Encounters of the Environmental Kind.* Anything goes, so have fun with it!

- The planning section provides a timeline of specific tasks to plan your camp from start to finish. The timeline offers a checklist and lots of detailed planning tips. We've also provided examples of enrollment forms, confirmation letters, a promotional flyer, and other support materials.

- Under each of the four main topics, we've provided more than enough activities so that you can choose specific activities to match participants' interests and abilities.

**Camper Notebook: The second section of the guide provides reproducible activity pages.**

- Reproducible Camper Notebook pages contain complete self-guided instructions and explanations so young investigators are empowered to do the activities individually or in groups.

- An accompanying CD-ROM containing all of the Camper Notebook pages in editable format allows you to change the procedures to meet your needs.

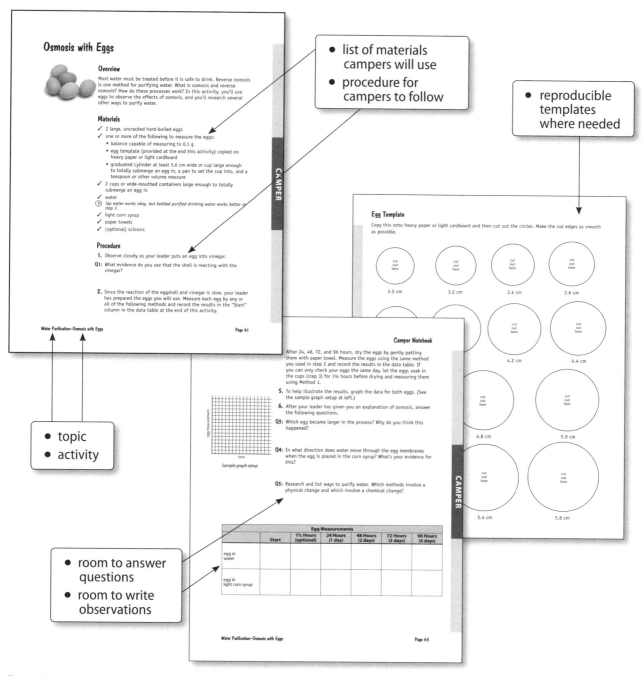

## Leader Guide: The third section of the guide provides leader resources for every camper activity.

- The Leader Guide for each activity provides suggestions to lead campers through the investigations. Important features include background information, introductory instructions, procedure notes and tips, wrap-up discussions, and answers to camper questions.

- brief summary of activity
- relevance of activity in context of unit

science background information
- review for experienced leader
- introduction for leader with less science background

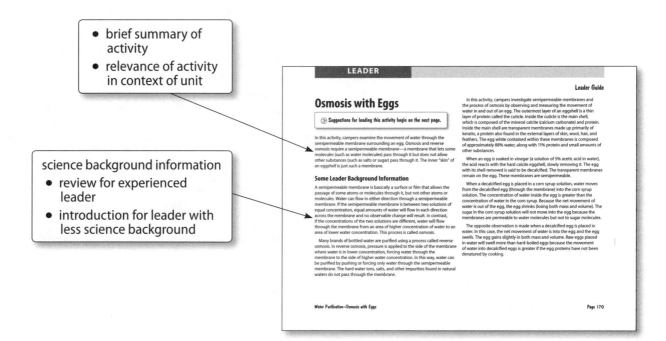

- small view of relevant camper page

- materials leader will need to get ready
- instructions for getting ready

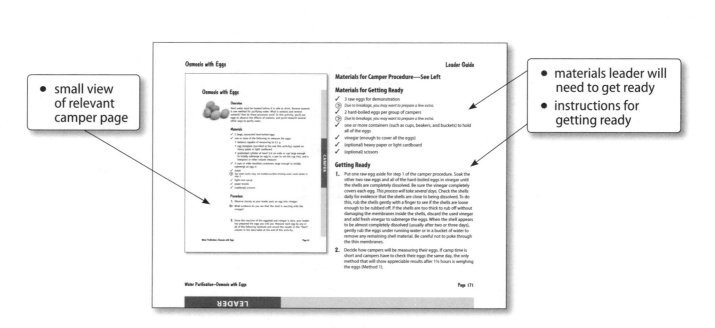

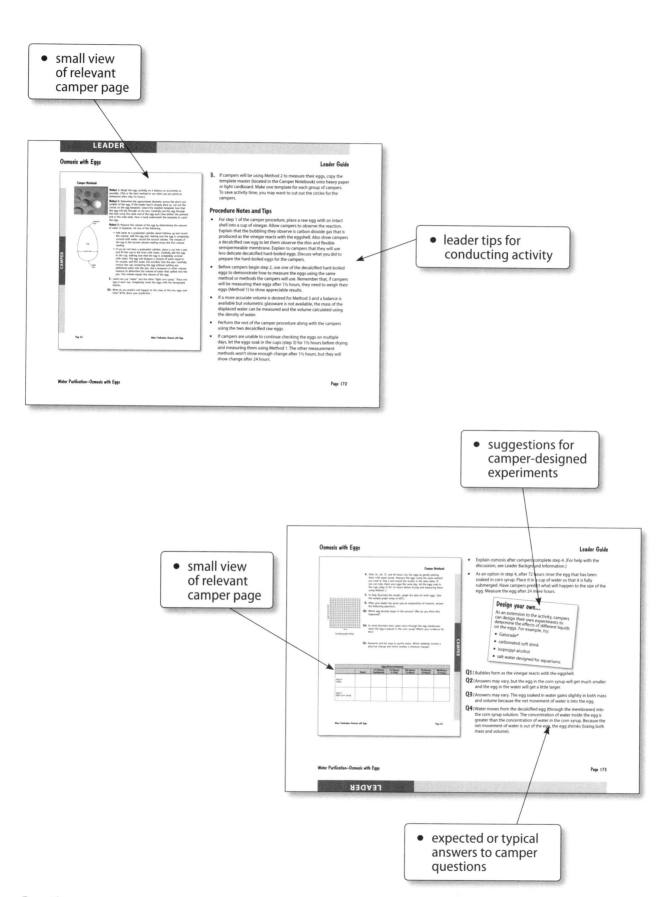

# PLANNING YOUR EVENT

This section includes Planning Made Easy and Example Materials.

PLANNING

# Planning Made Easy

One mark of a great science program is good planning. Use the following checklist to plan your event from start to finish. Use the detailed tips on the following pages as needed. Count down to the fun!

| | | | |
|---|---|---|---|
| **9 MONTHS AND COUNTING** | **Plan Early!** | ☐ Prepare a Budget<br>☐ Select Facilities<br>☐ Decide on Staffing | Check out pages 4–6 for details. |
| **6 MONTHS AND COUNTING** | **Finalize Plans!** | ☐ Confirm Staff Members<br>☐ Confirm Dates<br>☐ Confirm Facility Reservations | Check out page 7 for details. |
| **3 MONTHS AND COUNTING** | **Lock in Details!** | ☐ Recruit Participants<br>☐ Register Participants<br>☐ Develop a Schedule<br>☐ Make a Supply List | Check out pages 8–12 for details. |
| **1 MONTH AND COUNTING** | **Get the Goods!** | ☐ Gather and Shop for Supplies | Check out page 12 for details. |
| **2 WEEKS AND COUNTING** | **Final Contacts!** | ☐ Increase Community Awareness<br>☐ Contact the Facility<br>☐ Contact Staff<br>☐ Contact Campers | Check out page 13 for details. |
| **1 WEEK AND COUNTING** | **Final Prep!** | ☐ Final Setup<br>☐ Staff Training<br>☐ Prepare Name Tags and Materials | Check out page 14 for details. |
| **THE BIG EVENT HAS ARRIVED** | **Now Have Fun!** | ☐ Before Campers Arrive<br>☐ When Campers Arrive<br>☐ Winding Down | Check out pages 15–16 for details. |

Plan Early!

☐ **Prepare a Budget**

By carefully planning the budget, you save time and trouble later. The camp's budget depends primarily upon the number of campers and the camper fee you establish. If you are planning five half-day sessions, you may want to start your budget by estimating a camper fee per camper, with the goal of registering 30 campers. Remember that your expenses and income will change proportionally with the number of campers served.

It may be desirable to grant full scholarships to some participants (perhaps five or six per 30 campers). These scholarships help economically disadvantaged campers attend the camp and perhaps increase minority participation in the program. The scholarships can be granted to campers who receive full school lunch assistance. You may want to give assistance preferentially to minority campers and to girls. (It may seem odd to need to encourage girls; however, old stereotypes still persist.) Granting some scholarships to economically disadvantaged campers may help with obtaining grant support or other sponsorship of the camp. Grants and sponsorships can provide some breathing room in the budget.

One of the biggest items in the budget is staff. The required staff obviously depends upon the size of the camp. A good estimate has a small-group leader supervising groups of five to eight campers. Younger campers will require closer supervision in hands-on sessions.

If the director of the camp is a permanent faculty or staff member of your institution, it is possible to have the director position cost donated by your institution. Retired scientists, industrial scientists, or health professionals may volunteer their time as presenters.

The amount you budget for activity supplies should cover materials for the camp activities and take-home items, general camp supplies, and optional camp T-shirts (which serve to advertise the camp). See page 12 for tips on making a supply list.

If your camp is held at a school, university, or other community site, the space to run the camp may be donated. Otherwise, funds will be necessary to rent space.

| Budget Worksheet | |
|---|---|
| **Income** | |
| camper fees | $ |
| sponsorships and grants | $ |
| other income | $ |
| TOTAL INCOME | $ |
| **Expenses** | |
| director | $ |
| presenters | $ |
| small-group leaders/assistant | $ |
| activity supplies | $ |
| duplication costs | $ |
| flyers, postage, advertising | $ |
| use of facilities | $ |
| other expenses | $ |
| TOTAL EXPENSES | $ |

## ☐ Select Facilities

The location of your camp must be reserved well in advance. The ideal location for the camp involves multiple spaces, including rooms for activities and for check-in and gathering. At minimum, the check-in and gathering space could be a hallway, but a room will work better. The activity room is ideally a room with low tables and chairs, access to running water, and numerous sources of electricity. Many all-purpose rooms in museums and libraries have these features.

Some thought should also be given to accessibility of both the check-in/gathering and activity spaces. A location that is easy for campers to find from a parking lot or other drop-off location is convenient for both campers and their drivers. Ideally, the space will also allow access by and accommodation of campers and staff with disabilities.

If possible, you might consider facilities that include a large room capable of seating parents and siblings as well as campers. This large room allows parents and siblings to observe the opening and closing presentations and encourages at-home discussion about the camp activities. If the camp is expanded to 60 campers, it is especially helpful if there is one large presentation room and two activity rooms in addition to a place for check-in and snacks. Lastly, a staging area or room where supplies can be stored just before, during, and just after the camp is also useful.

After deciding on the ideal location for the camp, find out when it is available. If you are using the facility for several consecutive days, try to arrange to leave some materials set up overnight. In addition, reserving the facility for a few days before and after the camp for setup and cleanup is very helpful. From the times the facility is available, choose the preferred period and one or two alternate periods. If possible, make tentative reservations for all these times. The unneeded reservations can be cancelled once a final date for your camp is selected. Be sure to obtain written confirmation of your room reservations from the director of the facility.

## ☐ Decide on Staffing

The early planning stage is the right time to make several decisions on how the camp is to be staffed. Depending upon your organization, it may be possible to staff the camp with personnel from within, or you may need to recruit outside personnel. The following model allows the workload to be divided into manageable pieces; it also has some convenient hierarchy.

Our camp hierarchy is as follows: The camp is run by a director. The director is in charge of all aspects of the camp. The director supervises a number of presenters. It is convenient for each presenter to be in charge of one or more of the four major topics. Each presenter introduces the activities within their topic with engaging pre- and post-lab presentations. The presenter also assists the small-group leaders in performing and presenting the activities. In the activity part of the camp, each small-group leader is responsible for leading or guiding five to eight campers through the activities. (One possible way to make this hierarchy work with a group of five or six equally talented staff members might be to designate each small-group leader as also a presenter for one topic. The other staff members function as small-group leaders on the day featuring the presenter's topic. Any extra staff not assigned a topic to present can assist the other presenters as needed.)

At this time, make preliminary contact with people who may want to serve as staff members. The following page lists the duties of each staff member.

# WANTED

## Camp Director
The director will oversee all administrative details as described in this book, such as hiring staff, keeping records, and managing the budget. He or she will also assist with presentations if needed. The director works with the presenters to select activities and demonstrations. The director is also responsible for making sure that safety protocols of each activity and presentation are observed.

## Small-Group Leaders
The small-group leaders work with five to eight campers to help the campers successfully complete the hands-on activities. A small-group leader should be aware of each camper's progress and offer helpful comments or "what-if questions" as appropriate. The small-group leader's interest in participants, observations, results, and conclusions is essential to the camper's experience. Small-group leaders may be college or advanced high school students as well as teachers or volunteers from industry or the community. Small-group leaders may also help presenters prepare for activities and demonstrations. While teachers or scientists may bring more experience in their role as small-group leaders, college and advanced high school students often bring youth and enthusiasm to the groups. Having small-group leaders from diverse backgrounds enriches the camp experience.

## Presenters
The presenters are often teachers of physical and natural sciences, scientists, naturalists, or others with special expertise. During the camp day, a presenter typically performs demonstrations and introductory activities and guides the small-group leaders and campers through the hands-on activities. The presenters should be very familiar with the goals and activities of the camp. Campers in the programs often report that the presenters and small-group leaders are one of the best things about the camp experience.

## Assistant
An assistant who can run errands, look for misplaced materials, set up punch and cookies, and lend an extra hand when needed is indispensable during the camp. Presenters are often focused on the pre- and post-activity presentations and the small-group leaders are busy leading individual small groups, so the assistant is really necessary. Also, if a small-group leader becomes ill, the assistant can substitute for him or her.

## Finalize Plans!

| | |
|---|---|
| ☐ **Confirm Staff Members** | Contact potential presenters to find out their schedules. Presenter availability may determine the final choice of the dates. Arrange for the presenters to present topics in a logical order on consecutive days of the camp. Decide on the number of small-group leaders you will need and begin to secure small-group leader commitments once the final dates for your camp have been chosen. Be sure all presenters and small-group leaders are aware of the ages of the campers, so that presentations can be planned accordingly. If a planning meeting is not possible for the director and presenters, the director should let each presenter know what presentations are planned by other presenters. |
| ☐ **Confirm Dates** | Decide on the final dates of your camp and staff training session. The staff session should occur about one week before the camp. (See "1 Week and Counting" for staff training details.) |
| ☐ **Confirm Facility Reservations** | Once the director, presenters, and final dates are confirmed for the camp, be sure to confirm your reservation of the facility and cancel the reservations that will not be needed. If outside presenters, small-group leaders, and assistant are to be paid, meet with your institution's budget officer to find out how they should be paid. Since staff may need to visit the personnel office and complete several forms, be sure this is done well in advance of the camp. Since some funding sources require information on demographics and how many people your camp impacts directly and indirectly, you may wish to have your staff complete staff profiles. An example of a High School Student Camp Counselor Profile form is provided in Example Materials. |

 ## Lock in Details!

☐ **Recruit Participants**

Advertise the camp using your institution's existing online and print media (such as newsletters). Sister institutions, community centers, public libraries, schools, and other youth groups might also advertise your camp in their newsletters. To arrange for the articles to be published, first contact the editor or institutions where you would like an article to appear. Determine whether they would prefer to write the articles themselves from materials you supply or have you submit the article. (See an example press release in Example Materials.) If you write the article yourself, remember to send a camp brochure with it, as well as a cover letter inviting the institution or newspaper to contact you for further information. Local and regional newspapers might also carry short articles about the upcoming camps. The articles serve two purposes. Not only do they alert students and parents to the camp you are offering, but they also create good publicity for science education, your institution, and sponsors.

A good source of advertising is brochures, flyers, or posters that you produce and distribute. An example of a promotional flyer is included in Example Materials. Be sure to include the time and location of your specific camp and descriptions of some of the activities that will be offered. Also include all of the necessary registration information, including payment methods for fees. Use short, descriptive sentences throughout your flyer so that students will understand the camp topics. Have the flyer printed with colored ink or on colored paper, because students are more likely to take home brightly colored brochures. If you plan to offer scholarships, be sure the flyer mentions how to apply for a scholarship as well.

The number of flyers to copy and distribute depends upon your geographic area and the size of the camp you plan to run. About 50 brochures might be placed in each library in your area and 5–10 brochures mounted on community bulletin boards in your area. If the brochures disappear fast, distribute additional brochures. If you plan to distribute brochures through schools, plan on about one brochure per student in your target grades. Alternatively, ask that a school announcement be made about the camp. Students can be directed to pick up the brochure in a designated area. Some school districts require that flyers to be distributed in schools be approved by the Board of Education. Be sure to ask the school principal about their policy. Add about 100 to your total number of brochures to mail to people who request them and to keep in your files for future reference. Be sure that all staff members receive a copy of the brochure.

When distributing brochures to schools and other locations, try to contact the people who will be directly responsible for handing out or displaying the materials. (See a letter to coordinators example in Example Materials.) Briefly explain the goals of the camp. These individuals will be more likely to help if they believe the camp will be a good experience for students.

If you have previously hosted a camp and expect to do different activities in a second camp, send a brochure directly to all past campers. Encourage them to attend or to give the brochure to someone they think might like to attend.

## ☐ Register Participants

As the camp registrations arrive, record each camper's name, grade, age, birth date, T-shirt size, address, telephone number, and parent's or guardian's name on a master registration sheet. Identify campers who will be receiving scholarships. A computer spreadsheet program is helpful in organizing this information. Deposit checks from registered campers into the camp's account according to your institution's policies and procedures.

Within two weeks of receiving the registration forms, send letters of confirmation to the campers and their parents or guardians. (See Example Materials.) Send special letters to campers who applied for scholarships. (See Example Materials.)

In all letters to campers, be sure to include specific instructions on where and when campers should be dropped off and picked up each day. Enclose a map. (You might include a statement that balloons will be outside the designated entrance on the first day.) Enclose an emergency medical form and camp consent form for them to mail in before camp starts. Make sure they are aware a camper cannot participate without the emergency medical form. The camp consent form should ask parents and guardians to list the people who can pick up their children and also ask permission for photographing and videotaping their children. Since some outside funding sources require demographic information on campers, you may also need to have campers fill out a camper profile. See examples of all three forms in Example Materials. Enclosing a self-addressed, stamped envelope will encourage the parents to send the forms in early.

Remind parents that if they want to cancel their registration and receive a refund, they must notify you no later than one week before the camp begins. After you have accepted as many campers as you can accommodate, begin a second list. Record all applications even though the number may be far more than you could possibly include. You can then mail letters and brochures to these campers individually the following year, which is an important part of advertising.

For all registrations received after the camp is full, contact the parents or guardians to determine whether campers would rather be placed on the waiting list or withdraw their registrations. Be sure to tell parents how many children will be on the waiting list ahead of them, as well as how many cancellations you expect. (In general, only one or two campers will cancel.)

If parents or guardians of campers on the waiting list want to withdraw their registrations, return the uncashed checks immediately. If they want to be placed on the waiting list, be sure they understand that unless they hear from you, they should assume that their children were not admitted to the camp. Let them know that you will hold the checks uncashed until a spot opens up due to someone dropping out. At that time, you will contact the first person on the waiting list. Explain that after the first day of camp you will call them only if you can offer them a spot, and that if no spot opens up you will immediately return their checks.

## ☐ Develop a Schedule

The following scheduling scenario is for a five-day camp with three-hour daily sessions. (See an example five-day schedule in Example Materials.) You can modify your camp to include fewer activities in shorter sessions or limit the topics to one or more for a single session or weekend format.

Each three-hour session opens with a large group presentation. This presentation introduces the topics for the day, providing background and general knowledge on the topics. It is approximately 20–30 minutes long. Following the presentation, campers are dismissed into smaller groups to participate in hands-on activities. The hands-on activity period is approximately two hours long with a 15-minute break for juice and cookies. Campers can work with the same small-group leader both before and after the snack break, or they can work with one small-group leader before the break and another small-group leader after the break. Either way, the session wraps up with one large group closing presentation lasting approximately 20 minutes.

| Example Three-Hour Session |
| --- |
| opening presentation (20–30 minutes) |
| hands-on activities (1 hour) |
| snack break (15 minutes) |
| hands-on activities (1 hour) |
| closing presentation (20 minutes) |

The opening presentation sets the stage for the session. The topic is introduced with questions, surveys, role playing, and other attention-grabbing activities. Demonstrations are presented that illustrate the science concepts covered in the day's activities. Participants make observations, predictions, and conclusions about each demonstration. The general scientific principles that are important in the demonstrations are covered, as well as popular myths or assumptions about the risks inherent in a given environmental choice. Most importantly, each opening presentation is also an opportunity to discuss the safety procedures essential to working in the laboratory. When planning the opening presentations, you can modify the hands-on activities included in this book (see the example five-day schedule in Example Materials), show related videos, present local issues, or discuss news events.

The hands-on activity sessions are the heart of the science camp. While supervised in small groups, participants explore and discover for themselves. At least one activity involves making something to "take home."

A cookies-and-juice break, while not absolutely essential, makes the program friendly and hospitable. Participants also get to know each other in these social settings and enjoy their new friendships based on common interests.

The closing session is a final chance to review and expand upon what was discovered in the activity sessions. Do this presentation in a format similar to the introductory sessions by using demonstrations, role playing, and camper participation to reinforce the main concepts and challenges. Invite parents and siblings to join the closing session. A take-home activity may be distributed in the closing presentation to extend the fun and help campers describe their experience at home. Depending upon the level of camper participation in the opening and closing presentations, parents and siblings usually appreciate and learn along with the campers. These presentations also promote family discussion about camp activities.

It is important to note that your presenters may have additional insights or ideas to tie the topics to your specific locality. They may also suggest additional or different activities or attention-getting demonstrations.

Schedule the first day to begin 15–30 minutes earlier than the others (depending on the size of the camp). If you do not allow this extra time to check in campers and distribute materials, check-in may take up some of the time you had planned for your first demonstrations. Explain to the campers that you will begin each day with a demonstration and that if they are late they will miss it. You'll be surprised how prompt they are.

You may want to set up an activity on the first day that campers can observe all week. For example, campers can set up the "Osmosis with Eggs" on the first day and observe the changes of the eggs on subsequent days. The changes may be the most dramatic from the first day to the second day, but campers will be curious about what will happen if the samples remain in their solutions for even longer times.

Schedule two hands-on activities and two or three brief demonstrations for each day during the hands-on activity time. Be sure that one hands-on activity each day produces something that campers can take home. The take-home products of the camp (such as the soap, Cartesian diver with 1 ppm glitter, and lather print) are conversation starters for discussing the camp with family.

After drawing up a basic schedule, consider a few extra activities or extensions to the planned activities. These can be used in case an activity fails for any reason or in case some campers finish early. Activities not already planned for the camp or activities designated as take-home activities can be printed up and used during the camp if some of the planned activities conclude early.

Many activities can be done quite quickly if all of the physical and chemical changes are not observed, measured, and discussed. This should be discouraged. This camp is not a science race. Campers should work at a comfortable pace and be encouraged as developing scientists. The role of the small-group leader is often to look over a camper's shoulder and point out things they may have missed. Since the topics covered are camper-environment related, the scientific and societal implications are also worth noting and discussing. Both rushing campers through the activities and dragging the activities out and boring campers should be avoided. Achieving the right pace is also easier if the group size is small. While eighth-grade campers can work in a group of eight campers with one small-group leader, eighth-grade campers might also grow more as scientists with greater individual mentoring. A group size of six campers per small-group leader is often ideal.

☐ **Make a Supply List**

After selecting the activities and demonstrations for your camp, prepare a master list of the supplies. Make your list from the Materials section of the Camper Notebook and the Materials for Getting Ready section of the Leader Guide. Make sure to correctly indicate the quantity of each item, including extra supplies for demonstrations and in case of breakage. If you list the amount needed for either each camper or each group, as well as the amount for the entire camp, the amount needed for the entire camp can be adjusted as the number of applications to the camp increases. It is also helpful to list the proposed supplier for each item. Grouping all the supplies needed for each activity makes it easier to organize the materials once they arrive. A computer spreadsheet can be helpful. (There are two sides to handling the materials for the camp. Shopping is easier if all like items are grouped together, but performing the demonstrations and activities are easier if the supplies for individual activities are grouped together.)

Other items are needed besides the activity items. Each camper, presenter, and small-group leader will need goggles, pencil, paper, name tag, and folder. (Some camps also purchase T-shirts with the program logo and/or disposable lab coats for campers to keep.) Paper towels, soap, and first aid supplies must be available in every room you will use, and all program staff members must be prepared to deal with both small and large medical emergencies. You will also need cookies, punch, cups, and napkins for each day's snack break.

Pictures of individual campers participating in the camp are also nice take-home items. The pictures can be placed in a folder with the camp logo on the front. Campers also appreciate notes of encouragement and congratulations in these folders from the camp staff at the conclusion of the camp. Pictures can be taken with a digital camera and printed overnight. Another possibility is to purchase or borrow a camera with film that develops instantly. Make sure the camera has film (if needed), batteries, and a flash. Regardless of the media coverage that you might have arranged, you may want to take pictures for your own files and for future advertising. Be sure to get proper permission for any photos that you are keeping. (See an example of a camp consent form in Example Materials.)

 ## Get the Goods!

☐ **Gather and Shop for Supplies**

Locate items to be supplied by your institution and organize them in a convenient location. Make sure you have the quantities you need. Local merchants might also donate supplies. (If so, try to acknowledge them as local sponsors of the camp.) Place orders for supplies to be delivered. Nonperishable supplies to be purchased at local stores can be purchased about 2 to 3 weeks prior to the camp.

As the supplies arrive, check them off on your master list and separate them according to activity. Some people find it helpful to place each activity's materials in a separate bin (plastic or cardboard) or plastic bag. Doing so will help you verify that you have everything you need and will save time when you set up each activity during the camp week.

# Final Contacts!

| | |
|---|---|
| ☐ **Increase Community Awareness** | One or two weeks before the camp begins, you may want to contact local and regional newspapers to arrange for coverage. You might also contact news directors of local television stations. If you have something special planned for the last day of the camp, you might arrange for television coverage then. (If possible, do not schedule a visit from the media on the first day of camp, since check-in might be a bit hectic and isn't really representative of the camp experience itself.) Make an appointment for the reporter to meet you at the camp facility. Doing so will help to guarantee both that the reporter will show up and that you will be prepared to greet the reporter when he or she arrives.<br><br>Provide media representatives with printed information about the camp or with a press release. (See an example press release in Example Materials.) Media representatives will be interested in the background and purpose of the camp as well as in the number of campers and the activities the campers are doing. If safety conditions permit, invite media to visit the activity rooms and take pictures while the campers are performing activities. (Be sure that you have received camp consent forms from all campers being photographed.) Be prepared to explain the activities the campers are participating in as well as others they have recently completed. Consider whether to allow reporters to talk to the campers.<br><br>Do not forget that perhaps the best sources of publicity are the campers. Sending home activity laboratory sheets, cards with photos of the campers, take-home activities, and T-shirts will surely make an impression on the parents. Make sure that the materials campers take home reflect the quality of work you have put into planning the camp. While it is important that the materials you have prepared look professional, the camp take-home products should really be made by campers themselves and reflect how involved the campers are in the hands-on activities. The campers' natural excitement will also convince their parents that the camp was a worthwhile experience.<br><br>If you plan to offer camps in the future, consider videotaping the campers and small-group leaders in action and developing a presentation you can show to potential funding agencies. (Again, be sure that you have received camp consent forms from all campers being videotaped.) |
| ☐ **Contact the Facility** | About two weeks before the camp, confirm the reservations for the facility. Let facility staff know the times you will need the building or rooms unlocked, if applicable. |
| ☐ **Contact Staff** | Confirm with the presenters that they are still committed to their roles in the camp. If you asked them to gather the supplies they need, find out the status of this task. Let presenters know when and if they need to share final instructions about the hands-on activities with the small-group leaders. |
| ☐ **Contact Campers** | If it has been a while since you sent the letters of confirmation to the campers, you may want to send an e-mail, postcard, or letter reminding them about the details of the camp. |

# Final Prep!

### ☐ Final Setup

Group supplies according to the days they will be used. With a detailed schedule in hand, mentally run through each day's activities to be sure you have everything you need. Do not rush—give yourself time to think. You may want to ask someone to work with you to help you remember everything. Working with an assistant as you do this also allows the assistant to know what will be needed and where the supplies are.

Make up all solutions called for in the activities and place them in small well-labeled bottles. Include name and formula on the labels. Include labeled spoons for all powders and crystals that the campers will measure—consider using colored tape to coordinate the containers and the spoons so that campers do not contaminate the chemicals by mixing up the spoons.

Prepare an area in each activity room (away from the campers' work area) where the presenters and small-group leaders can store supplies.

### ☐ Staff Training

About one week prior to the camp, try to hold a half-day formal preparation session. Ideally, this meeting should be scheduled in the facility that has been reserved. The director, presenters, and small-group leaders should meet to discuss the camp presentations and practice the activities. Running through the activities will serve as a final check that everything is present, and will let the small-group leaders become very familiar with the hands-on activities. The daily schedule, camp philosophy, and any unusual procedures should also be reviewed.

Instruct all staff members to arrive one hour before the camp begins each day and to bring a notebook to record their observations. Let them know that they should also allow about 30 minutes at the end of each day for cleanup, reorganization, and journal writing. In their journals, they should take notes about how the activities go and their ideas for improvement. You will later review these notes to determine what you may want to do differently in the future.

### ☐ Prepare Name Tags and Materials

Organize the registered children into groups of five to eight campers, depending on the number of small-group leaders you have. You may want to create groups with mixed ages and genders. Make name tags for the campers and small-group leaders that indicate the groups in which they belong. For example, give each group a color designation and prepare color-coded, durable name tags for the campers and small-group leaders. (Use either colored paper or colored markers of the group colors to make the name tags.) Be aware that color-blind campers may need extra help in identifying what group they are in.

Also make name tags for all staff members.

To help the first-day check-in process go as smoothly and quickly as possible, consider pinning camper name tags to their T-shirts if you plan on distributing the shirts on the first day. The shirt can also be placed in a folder with the name tag, pencil, blank paper for notes, and first-day activity sheets.

# Now Have Fun!

### ☐ Before Campers Arrive

The day you've planned for finally arrives! With all the preparations made, you are ready to begin the camp. Dress for the occasion—wear comfortable but somewhat stylish clothes; you are a positive role model for the campers. You are about to deliver a unique experience to the campers. Despite thousands of hours of schooling, most campers have not worked with non-family member adult experts in a field of mutual interest. Years later, whether or not campers have pursued careers in education, science, or technology, they will remember that science professionals were encouraging and interested in them personally.

You can place balloons and welcome signs outside the entrance to be used at your facility. If your check-in location is hard to find, use additional wall signs to assist campers in finding the location.

When staff members arrive an hour before the camp begins, give them a brief overview of the day's activities and answer their questions. Also give the small-group leaders a list of the campers who will be in their groups. Supervise as presenters move supplies to the activity or demonstration sites. Make sure that all supplies are ready to use (for example, paper figures cut out, solutions made, and chemicals pre-measured, if applicable).

Fifteen minutes before you expect campers to arrive, give a staff person a copy of the registration list and a highlighter pen to check off camper names as they arrive. Give a second staff member the campers' name tags, folders, and T-shirts (if applicable).

### ☐ When Campers Arrive

Make available extra emergency medical forms, camp consent forms, and camper profiles for the parents or guardians to fill out if they did not mail the forms back before the start of camp. *No camper should be admitted to the camp unless he or she has submitted an emergency medical form and a camper profile (if required by the outside funding source).* Do not allow campers to bring an unregistered friend or relative to camp (even for a day) because it is not fair to the registered campers who have paid for their share of the supplies. Perhaps more importantly, you would not have emergency medical forms for these campers in case an emergency occurred.

If a registered camper does not show up for the first day of camp, you might call his or her parents after one hour to determine whether he or she is still planning to attend. If spots open up, call parents whose children's names are on the waiting list to ask whether they would like to attend the remaining camp sessions. If you plan to charge a prorated fee, let them know.

Campers should leave their name tags and folders at your site from day to day. If you hand out laboratory coats to campers, ask the campers to affix their name tags to the lab coats. At the end of each day, have the campers roll up their goggles in their lab coats and give their lab coats to their small-group leaders. The small-group leaders can then make sure each camper gets his or her own goggles each day by looking at the name tag on the lab coat. Alternatively, at the end of the activity session, ask campers to pin their name tag to or around the goggles strap. The goggles with a name tag can be placed in a bin at the end of the activity session.

Be prepared to send home any campers who cause disciplinary problems by calling the parents and asking them to come pick up the camper. Usually, disciplinary problems are not an issue at these camps because the campers are too involved in the activities to get into trouble.

During the activities, make sure that all campers are wearing goggles at all times. Even if the activity they are performing at a given time is not hazardous, they should adopt the habit of always wearing goggles when they are in the activity room.

Do not forget to take pictures of individual campers and groups sometime before the last day. Try to get good shots of all the campers to incorporate into their cards or folders. Give the photographs to the small-group leaders to distribute to their campers on the last camp day. Alternatively, distribute them at the last closing presentation.

Send a note about halfway through the camp to invite parents and guardians to attend the final closing demonstration, if this is appropriate or possible.

Arranging for groups of campers to work at their own pace and finish a set of activities at the same time is impossible. The emphasis in this camp is on quality rather than quantity. If campers are happy with their work and complete the activities that involve the take-home projects, some of the uncompleted activities might be completed at home. Special instructions on individual topics and activities are included in this guide.

Campers do not necessarily need to leave the space as clean as they found it, but cleaning as one works is part of working responsibly. Disposing of waste properly is an important part of maintaining a healthy environment. Campers need to be coached and helped in cleaning up the activity room.

## ☐ Winding Down

Camper input can be useful to help you improve your camp. Consider asking campers to fill out camper evaluation forms (see Example Materials) to evaluate their camp experience.

During the final session, greet parents as they pick up their campers. Let the parents and campers know which agencies and/or sponsors helped with funding for the camp. If possible, provide names and addresses of people to whom they can send notes if they would like to express their support for the program and help ensure that it will be funded again next year. Collect thank you notes from campers who received scholarships or give them the names and addresses of the funding agencies or sponsors and remind them to write thank you letters expressing their appreciation.

Sign the folders or pictures of campers in the camp. Point out how they have grown in their science understanding. Encourage small-group leaders to personally say goodbye to each of the campers in their groups. Thank the staff publicly for their contributions to the camp. As the campers leave, thank them for participating in the camp and invite them back for next year.

After the campers and parents leave, thank your staff again. Make sure you have their addresses so that you can send them paychecks, thank-you cards, and other materials. If high school or college students have performed exceptionally, offer to be a job reference if they need one in the future.

Ask the staff to submit their observations, journal comments, and suggestions for the future. Also make notes of changes or new plans that you may want to implement next time.

Consider having some kind of reception/recognition for your staff. The camp could not have happened without them. Try to leave them with a very positive memory of the experience—make sure they know their efforts were appreciated.

# Example Materials

The documents on the following pages are examples of forms, letters, and flyers that can be adapted for your camp.

- ☐ Five-Day Schedule
- ☐ High School Student Camp Counselor Profile
- ☐ Press Release
- ☐ Promotional Flyer
- ☐ Letter to Coordinators
- ☐ Confirmation Letter for Applicants
- ☐ Scholarship Letter for Applicants
- ☐ Camp Emergency Medical Authorization
- ☐ Camp Consent Form
- ☐ Camper Profile
- ☐ Camper Evaluation Form

# Example of Five-Day Schedule

## Science Is Fun Camp

| | Monday<br>Clean Hands Day | Tuesday<br>Water Day | Wednesday<br>Ecology Day | Thursday<br>Air Day | Friday<br>UV Day |
|---|---|---|---|---|---|
| **Check-In**<br>8:30–9:00 AM | hand out name tags, folders, and shirts | | | | |
| **Opening**<br>9:00 AM<br>Room 115 | • "Watching Granny Smith Rot" take-home activity as demo | • "Look at 1 PPM" as demo (do on an overhead projector)<br>• check apples | • discuss ways to purify water or do "PUR" demo<br>• check apples and crackers | • "Balloon Challenge"<br>• "Search for 1 PPM"<br>• check apples and crackers | • "Visible Light Challenge"<br>• check apples and crackers |
| **Lesson A**<br>9:25 AM<br>Room 120 | • "Why Soap?"<br>• "Design Your Own Soap" | • "Back to Its Elements"<br>• "Distinguishing Water Samples" | • "Investigating Mineral Content"<br>• "Hydrophobic Art"<br>• check crystals | • "Humidity Detector"<br>• "How Much Oxygen Is in Air?" | • "How Sensitive Is Your Skin?"<br>• "Make and Test Lip Balm" |
| **Break**<br>10:20–10:35 AM<br>Lobby | snack break | snack break | snack break | snack break | snack break |
| **Lesson B**<br>10:35 AM<br>Room 117 | • "DNA from Strawberries"<br>• "Colorful Lather Printing" | • "Osmosis with Eggs"<br>• "Water Taste Test" | • check eggs<br>• "How Hard Is Your Water?"<br>• "Purify Surface Water" | • "Pour a Gas"<br>• "Pour More Gas" | • "Cover Up, Screen, or Block?"<br>• "Sunning Straws"<br>• make UV bracelets per "UV Detective Challenge" take-home activity |
| **Closing**<br>11:40 AM<br>Room 115 | • "Surfactants" | • "Growing Mold" take-home activity as demo | • "Removal of Iodine from Water" as demo | • "Stirring Indoor Air"<br>• "Trapping Particulates" | • fill out evaluations<br>• hand out pictures and cards |
| **Take-Home Materials** | • soap<br>• lather print<br>• "How Clean Is Your Clan?" take-home activity | • "What's Really in the Bottle?" take-home activity | • hydrophobic art<br>• "The Amazing Water Maze" take-home activity | • humidity flowers | • bracelet<br>• lip balm<br>• "UV Detective Challenge" take-home activity |

# High School Student Camp Counselor Profile

Dear Counselors: Please complete the following items. This information is needed for a federal report about the project in which you participated. All information will be kept strictly confidential.

Name _____ Today's date _____

Address _____

City, state, zip _____

Parent/guardian's name _____

Home phone _____ Work phone _____

Are you a  ☐ boy  ☐ girl     Birthday (month/day/year) _____

Your age now _____

What will your grade be in September? _____

Name of school you will be attending in September _____

Street address of school (if you know it) _____

City, state, zip _____

School phone (if you know it) _____

Superintendent's name (if you know it) _____

Is your school a public or non-public school?     ☐ public  ☐ private

What is the name of your school district?_____

What county do you live in? _____

Are you:  ☐ Caucasian  ☐ African American  ☐ Asian  ☐ Hispanic  ☐ Native American
          ☐ other

Please put a check mark if you are from the following groups/areas:
    ☐ gifted and talented     ☐ migrant

Economically disadvantaged (eligible for free or reduced-cost lunch)?  ☐ yes  ☐ no

Limited English proficiency?  ☐ yes  ☐ no        Handicapped?  ☐ yes  ☐ no

In which of the following types of areas is your school located: (check one)

    ☐ urban (community with population of 50,000 or larger)

    ☐ suburban (community with population of 2,500 to 50,000)

    ☐ sparsely populated (community with population less than 2,500)

# Example of Press Release

Contact: Suzy Leonard
telephone: 555/555-5555 ext. 123
e-mail: suzyleonard@terrificscience.org

FOR IMMEDIATE RELEASE

SCIENCE IS FUN SUMMER SCIENCE EVENTS

One hundred twenty local children in grades 3–7 are participating in *Science Is Fun* Summer Science Events this summer at Miami University in Hamilton and Middletown.

*Science Is Fun* Summer Science Events make science accessible to these campers by providing exciting, hands-on science experiences that challenge the campers to explore aspects of the scientific method. During the event week, campers engage in a variety of scientific experiments and activities related to disease control, water purification, healthy air, and healthy skin. Among other things, participants taste-test bottled water, try different hand-washing techniques, analyze sun screens, and purify water.

*Science Is Fun* is built on the premise that a basic understanding of scientific principles and methods is fundamental to the education of all children. The *Science Is Fun* program is sponsored by the Ohio Board of Regents through the Dwight D. Eisenhower Mathematics and Science Education Program and Miami University, with support from AK Steel, the Cincinnati Section of the American Chemical Society, and other private businesses, industries, and foundations in southwestern Ohio.

###

If you want more information about this event, please call Suzy Leonard at 555/555-5555 ext. 123.

# Example of Promotional Flyer (Page 1)

Miami University Middletown
4200 E. University Blvd.
Middletown, OH 45042-3497

Stamp

Darryl Jones

1234 Main Street

Middletown, OH 45042

# The Science Times

**SPECIAL EDITION!**  2008

## Check Out Science Is Fun Camp!

This camp has something for everyone. Delve into the mysteries of health and environmental science while doing cool activities and experiments. Explore what blowing bubbles has to do with disease control, discover how sensitive your skin is, use eggs to learn about water purification, and lots of other fascinating stuff. Intended for curious campers who want to do lots of different science activities.

- Camp dates: June 23–27
- Camp location: Miami Middletown Campus
- Camp time: 9:00 AM–Noon
- For students in grades 3–7 as of September
- Registration fee: $85
  (Applicants who receive full school lunch assistance may qualify for a limited number of scholarships. Contact us for details.)

**Registration fee includes instructional fees, camp supplies, take-home items, and a T-shirt.**

# Example of Promotional Flyer (Page 2)

## 2008 Summer Science Is Fun Camp

Registrations processed on a first-come, first-served basis. Sign up early to be sure of a space. Confirmation of your child's enrollment will be mailed to you prior to the start of camp. If it is necessary to cancel your registration, you must do so at least one week before the first day of camp. After that date, your money will not be refunded.

Be sure to mark your child's T-shirt size on the registration form; otherwise, he or she will not receive one. Because of printing limitations, *only adult-sized T-shirts are available.*

Camp information and applications are also available at www.terrificscience.org/parentskids/.

---

### Camp Registration Form — Hurry, Space Is Limited!

*Please print clearly and use a separate form for each camper.*

Child's name _____ Age _____ Birthdate _____

Address _____ Grade in fall 2008 _____

City _____ State _____ Zip _____ ❏ Male  ❏ Female

Parent/guardian's name _____ Student's school _____

Home phone _____ Work phone _____ E-mail _____

Method of payment:  ❏ Check   ❏ VISA   ❏ MasterCard

Card # _____ Exp. Date: _____

Parent/guardian signature _____

| Camp | Dates | Location | Fee |
|---|---|---|---|
| 2008 Summer Camp | June 23–27 | Middletown | $85 |
| For office use only   Received_____ | | Total amount: | |
| Forms sent_____   Check #_____ | | $_____ | |

**Adult T-Shirt Size: (Circle One)**

| S | M | L | XL |
|---|---|---|---|

Make your registration check payable to
Miami University.

Mail it with this form to
**Center for Chemistry Education
Miami University Middletown
4200 E. University Blvd.
Middletown, OH 45042-3497**

For more information, call Kitty at
555/555-5555

# Example of Letter to Coordinators

To: Building Principals, Science Coordinators, and Educational Resource Persons in Butler County

From: Mickey Sarquis, Program Director

Subject: *Science Is Fun* Summer Science Events

Summer is almost here and *Science Is Fun* Summer Science Events are just around the corner!

*Science Is Fun* Summer Science Events provide children in grades 3–7 with hands-on experiences in a variety of scientific disciplines. Participants will don lab coats and goggles and head into the laboratory to make soap, use a humidity detector, taste-test bottled water, and conduct many other experiments.

*Science Is Fun* participants will conduct experiments to learn how soaps and detergents work, how much oxygen is in air, how to purify water, and much more.

Events fill up fast so please encourage your students to apply early. We are offering free tuition to a limited number of students from low-income families. We hope you will encourage underprivileged students in your service area to apply for free tuition so that they, too, may benefit from this experience.

Enclosed are brochures describing the summer events, including information on dates, locations, fees, and activities. The brochures also contain applications for admission and free tuition. Please feel free to make photocopies of the brochure and/or the application(s) if necessary.

Thank you for helping us publicize this program; the response in years past has been outstanding and we look forward to again introducing young people to the fun of doing science. If you have any questions regarding the *Science is Fun* program, please contact Suzy Leonard, *Science Is Fun* Program Manager, at 555/555-5555 ext. 123.

# Example of Confirmation Letter for Applicants

May 28, 2008

To the parents/guardians of **Skylar McFee**:

Thank you for enrolling your child in the science camp. Your registration fee has been received and your child's name has been added to the list of campers.

Campers will meet **Monday through Friday, June 23–27 in room 115 Johnston Hall**. Camp will begin each day at **9:00 AM.**

**On June 23:**
Please come to the **Johnston Hall Lobby at 8:30 AM** to pick up your check-in materials. Parking is available in the south parking lot by Johnston Hall. (See enclosed map.)

You may pick up your child each day at noon in room 115 Johnston Hall.

Please complete the enclosed *Camp Emergency Medical Authorization* and the *Camp Consent Form* and return them as soon as possible to:

> **Liz Tyler**
> **Miami University Middletown**
> **4200 E. University Blvd.**
> **Middletown, OH 45042**

If we do not receive both forms **before the first day of camp,** your child will not be permitted to attend camp. Should it become necessary to cancel your registration, you must do so **at least one week** before the first day of camp. After that date (June 16), we will not be able to refund the registration fee.

If you have any questions, please feel free to call Liz at 555/555-5555.

Sincerely,


Suzy Leonard
Program Manager

# Example of Scholarship Letter for Applicants

May 28, 2008

To the parents/guardians of **Shawna Greene**:

We are pleased to award your child a full scholarship to attend the Miami University summer camp. This scholarship, sponsored by the **Middletown Community Foundation**, will cover all camp expenses for your child.

It will be your responsibility to see that your child has transportation to and from camp. Campers will meet **Monday through Friday, June 23–27,** at the Middletown campus. Camp will begin at **9:00 AM**. On June 23, please come to the **Johnston Hall Lobby at 8:30 AM** to pick up your check-in materials. You may pick up your child at noon in room 115 Johnston Hall.

Please complete the enclosed *Camp Emergency Medical Authorization* and the *Camp Consent Form* and return them as soon as possible to:

**Liz Tyler**
**Miami University Middletown**
**4200 E. University Blvd.**
**Middletown, OH 45042**

In addition, **please have your child write a thank you letter to the Middletown Community Foundation** and return it with your forms. If your child is too young to write the letter, a parent may do it. Your child can then draw a picture and print their name on it. This will be forwarded to the Foundation. We must receive the forms and the thank you note **before the first day of camp**. If we do not receive all these before the camp session begins, we reserve the right to offer this scholarship to another child. A return envelope is enclosed.

This scholarship is **not** transferable. Should you find that your child is not able to attend the camp, **please** call Liz at 555/555-5555 **at least one week** in advance so we may offer the scholarship to another child.

If you have any questions, please feel free to call Liz.

Sincerely,


Suzy Leonard
Program Manager

# MIAMI UNIVERSITY MIDDLETOWN
# Camp Emergency Medical Authorization

_____  _____
Child's Name                      Child's School District

_____  _____
Child's Address                   Child's Home Telephone Number

**PURPOSE:** To enable parents to authorize treatment for children who become ill or injured while under school authority, when parents cannot be reached.

## PART 1—GRANT OF CONSENT

In the event reasonable attempts to contact me at _____(phone number) or _____(other parent/guardian) at _____(phone number) have been unsuccessful, **I hereby give my consent** for:

(1) The administration of any treatment deemed necessary by Dr. _____ (preferred physician) at _____ (physician's phone number) or Dr. _____ (preferred dentist) at _____ (dentist's phone number) or in the event the designated preferred practitioner is not available, by another licensed physician or dentist.

(2) The transfer of the child to _____ (preferred hospital) or any other hospital reasonably accessible.

This authorization does not cover major surgery unless the medical opinions of two other licensed physicians or dentists, concurring the necessity of such surgery, are obtained before the surgery is performed.

Facts concerning the child's medical history including allergies, medications being taken and any physical impairments or major disabilities to which our camp personnel or a physician should be alerted:_____

_____  _____
Signature of Parent/Guardian      Date

_____
Address (if different from child's stated above)

### PLEASE DO NOT COMPLETE PART 2 IF YOU HAVE COMPLETED PART 1

## PART 2—REFUSAL TO CONSENT

**I do not give my consent** for emergency medical treatment of my child. In the event of illness or injury requiring emergency treatment, I wish the University authorities to take no action or to:

_____

_____  _____
Signature of Parent/Guardian      Date

_____
Address (if different from child's stated above)

# MIAMI UNIVERSITY MIDDLETOWN
## Camp Consent Form

_____
Name of child

### PERMISSION TO PICK UP MY CHILD

While attending camp at Miami University Middletown, I hereby give permission for my child to be picked up by the following person(s):

(Please limit to three.)

1) _____

2) _____

3) _____

Children will not be permitted to leave the room unescorted.
A photo ID may be requested by the staff for verification of identity.

_____     _____
Parent/Guardian signature                                              Date

---

### PERMISSION FOR PUBLICATION

I hereby give permission to Terrific Science Programs to photograph and videotape my child during camp sessions and to publish the images of my child in materials used for advertisement or promotion of future programs.

_____     _____
Parent/Guardian signature                                              Date

# Camper Profile

Dear Campers: We need to know some things about each camper. Please answer these questions. We promise we won't tell anyone your answers. Your parents may help you.

Name _____ Today's date _____

Address _____

City, state, zip _____

Parent/guardian's name _____

Home phone _____ Work phone _____

Are you a  ☐ boy  ☐ girl      Birthday (month/day/year) _____

Your age now _____

What will your grade be in September? _____

Name of school you will be attending in September _____

Street address of school (if you know it) _____

City, state, zip _____

School phone (if you know it) _____

Your teacher's name (if you know it) _____

Superintendent's name (if you know it) _____

Is your school a public or non-public school?    ☐ public  ☐ private

What is the name of your school district? _____

What county do you live in? _____

Are you:    ☐ Caucasian  ☐ African American  ☐ Asian  ☐ Hispanic  ☐ Native American
            ☐ other

Please put a check mark if you are from the following groups/areas:
            ☐ gifted and talented        ☐ migrant

Economically disadvantaged (eligible for free or reduced-cost lunch)?  ☐ yes  ☐ no

Limited English proficiency?  ☐ yes  ☐ no        Handicapped?  ☐ yes  ☐ no

In which of the following types of areas is your school located: (check one)

　　　☐ urban (community with population of 50,000 or larger)

　　　☐ suburban (community with population of 2,500 to 50,000)

　　　☐ sparsely populated (community with population less than 2,500)

# Camper Evaluation Form

Your Age:_____  Your Grade: _____  Date: _____

1. Did you enjoy this camp?

2. Would you recommend this science camp to your friends?

3. Do you like hands-on science?

4. What was your favorite activity or activities?

5. What was your least favorite activity?

6. Were any of the camp sessions too difficult for you? Which ones?

7. Which take-home activity did you like best?

8. Did you share any camp or take-home activities with your family? Which ones?

9. Is science one of your favorite subjects in school?

10. Was it easy for you to understand the guest scientists?

11. Would you like to attend a similar camp next summer?

12. List the best things about this camp.

_____
_____
_____
_____
_____
_____
_____
_____

# PLANNING

# CAMPER NOTEBOOK

Leader Guide notes follow this section.

CAMPER

CAMPER

# TOPIC 1

## DISEASE CONTROL IS IN YOUR HANDS

### List of Activities

- ✓ Why Soap? .................................................................................................... 34
  *Try different hand-washing methods to learn which one works best at removing germs.*

- ✓ Design Your Own Soap ................................................................................ 37
  *Choose additives such as color, fragrance, and decorative objects to design your own soap.*

- ✓ DNA from Strawberries ............................................................................... 39
  *Observe how soaps and detergents affect the oil-like cell walls of strawberry cells.*

- ✓ Colorful Lather Printing .............................................................................. 42
  *Learn about the concepts of polarity and hydrophobicity while making fun marbling patterns on paper.*

- ✓ Surfactants ................................................................................................... 44
  *Learn how soaps and detergents work to remove dirt and germs from the skin.*

- ✓ Blowing Bubbles .......................................................................................... 47
  *Play with bubbles and learn details about their structure.*

- ✓ Take-Home Activity: Watching Granny Smith Rot ................................... 49
  *Discover the importance of hand washing before handling food.*

- ✓ Take-Home Activity: How Clean Is Your Clan? ......................................... 52
  *Observe how well your friends and family wash their hands.*

# Why Soap?

## Overview

How do disease-causing germs get on your hands and how can you get them off? In this activity, you will learn how diseases can spread. You will also experiment with different hand-washing methods.

## Materials

- ✓ glitter mixture to represent infection (provided by leader)
- ✓ paper towels
- ✓ access to sink with warm water or pitchers of warm water plus a bucket for waste water
- ✓ hand soap
- ✓ stopwatch or timepiece with a second hand
- ✓ hair dryer

## Procedure

1. Participate in a group introduction as directed by your leader.

2. Examine everyone's hands to discover who has been "infected." You may need to use a special light.

**Q1:** How did the infection get passed around? How many people became infected?

3. Work with a friend to test the various methods of removing germs from hands listed in the data table at the end of this activity. One person should be the hand washer and the other person should be the timekeeper and data recorder.

4. Squeeze a pea-sized portion of glitter mixture in the hand of the hand washer and have her or him rub the hands together so the mixture thoroughly covers the front and back of each hand. The glitter mixture represents germs.

5. Have the hand washer try the first method listed in the data table. Then, the data recorder should shade the hand diagrams to show where glitter and oily sheen remains on the palm and back of each hand.

## Camper Notebook

**6.** Repeat steps 4 and 5 for each method in the data table. Be sure the hand washer reapplies the glitter mixture just as thoroughly before trying each method. After trying all of the methods, rank each method's effectiveness from the most effective at removing germs (1) to the least effective at removing germs (6).

**Q2:** Which hand-washing technique do you normally use?


**Q3:** Based on the results of this activity, which hand-washing method is the most effective? Which is the least effective?

## Camper Notebook

| Method | Results | Observations |
|---|---|---|
| Wipe with a dry paper towel for 5 seconds (don't use soap and water). | palm    back | |
| Rinse with plain water for 5 seconds (but don't dry hands). | palm    back | |
| Rinse with plain water for 5 seconds and dry with a paper towel. | palm    back | |
| Wash with soap for 5 seconds, rinse with water, and dry with a paper towel. | palm    back | |
| Wash with soap for 5 seconds, rinse with water, and dry with a hair dryer. | palm    back | |
| Thoroughly wash with soap for 20 seconds while hands are not under faucet, rinse with water, and dry with a paper towel. | palm    back | |

# Design Your Own Soap

## Overview

Soap casting is currently a popular craft. Have you ever tried it? In this activity, you get to design your own soap by choosing additives such as color, fragrance, and decorative objects.

## Materials

- ✓ about ½ cup (125 mL) glycerin soap base (such as Neutrogena® or another, less expensive brand)
- ✓ knife (use with adult supervision)
- ✓ cutting board
- ✓ one of the following sets of materials to melt the glycerin soap:
    - microwave-safe container, plastic food wrap, and microwave
    - pan and hot plate or stove
- ✓ wooden chopstick or wooden spoon
- ✓ cosmetic-grade soap additives (such as scent, color, oils, and solids)
- ✓ materials to measure additives (such as droppers, measuring spoons, and cups)
- ✓ flexible mold (such as soapmaking mold, disposable paper or plastic cup, or small plastic bag)
- ✓ (optional) 70% isopropyl rubbing alcohol in spray bottle
- ✓ (optional) access to a refrigerator

## Procedure

1. With adult supervision, carefully cut approximately ½ cup (125 mL) glycerin soap base into small pieces measuring no bigger than 1 inch x 1 inch (3 cm x 3 cm).

2. Follow one of these two methods to melt the glycerin soap pieces:
    - Microwave method: Place the glycerin pieces in a microwave-safe container. Cover the container with plastic wrap (to prevent loss of moisture) and microwave in 10-second intervals, stirring in between with a wooden chopstick or wooden spoon. (Remove from the microwave after melted, usually after about 30 seconds.)

Disease Control Is in Your Hands—Design Your Own Soap

## Camper Notebook

- Stove top or hot plate method: Place the glycerin pieces in a pan. Carefully heat the pan while stirring with a wooden chopstick or wooden spoon. Remove the pan from the heat when all of the glycerin soap is melted.

3. Decide on the additives you want to use in your soap. Use the wooden chopstick or spoon to stir the additives into the melted soap according to the directions provided in the table below.

4. Pour the melted glycerin-additives mixture into the flexible mold.
   ☞ *If bubbles form on the surface of the melted glycerin, you can lightly spray the surface with 70% isopropyl rubbing alcohol.*

5. Allow the soap to cool to room temperature. (This may take a couple of hours.) Placing the mold in the refrigerator or freezer will speed up the cooling process, but cooling the soap much past room temperature may make it too hard and flaky.

6. Tap and flex the mold to remove the hardened soap. Bag and cup molds can be cut away or torn. Store your soap in plastic food wrap to prevent loss of moisture until you are ready to use it.

| \multicolumn{3}{c}{**Cosmetic-Grade Soap Additives**} | | |
|---|---|---|
| **Additive** | **Examples** | **Instructions** |
| scent | soapmaking fragrance | Add about ¼ teaspoon (1.25 mL) commercial soapmaking fragrance. With milder scents, add more fragrance with a dropper until the desired scent is achieved. |
| | body mist | Add 1 tablespoon (15 mL) body mist to the melted glycerin. Adjust the quantity until the desired scent is achieved. |
| | perfume | Allow the melted glycerin to partially cool before adding alcohol-based perfumes because the fragrance will evaporate quickly if the glycerin is too hot. |
| color | food color | Add a drop or two of food color. (The color won't come off on hands or towels when the soap is used.) |
| | soapmaking colorants | Follow instructions on the package. |
| oil | coconut, sweet almond, refined olive, avocado, jojoba, or lanolin | Add two or three drops of oil. |
| solid | dried flowers and herbs, grated citrus peel, ultra fine cosmetic grade glitter, or small plastic toys | Pour a layer of glycerin into the mold, letting the glycerin begin to harden. Add the solids, and pour another layer of glycerin on top. Fresh flowers and other plant materials may require preservative to prevent mold growth in the soap. When placed in soap, some fresh flowers such as rose petals and lavender may turn brown with age. |

# DNA from Strawberries

## Overview

Today, we know that many diseases are spread by germs people carry on their hands. We know that soap is important when washing our hands, but how does soap work to clean our hands? In this activity, strawberry cells with oil-like cell walls are related to oily and greasy dirt. You will discover what different soaps and detergents do to the strawberry cell walls.

## Materials

- ✓ 2 small zipper-type plastic bags (Freezer bags are preferred to sandwich bags because they are thicker. Do not use bags with bottom pleats.)
- ✓ 2 ripe strawberries
- ✓ 6 pinches table salt (sodium chloride, NaCl)
- ✓ 3 small cups (such as disposable bathroom cups or measuring cups that come with liquid medicines)
- ✓ water
- ✓ teaspoon or 10-mL graduated cylinder
- ✓ soap- or detergent-based cleaner, such as Ivory® bar soap, liquid dishwashing detergent, or shampoo without conditioner
- ✓ laboratory spatula or butter knife and measuring spoon (if using Ivory bar soap)
- ✓ dropper (if using liquid dishwashing detergent or shampoo)
- ✓ 4 test tubes
- ✓ 60° angle funnel
- ✓ 2 filter papers or cone-type coffee filters
- ✓ ice bath
- ✓ ice-cold 99% isopropyl alcohol (prepared by leader)
- ✓ 2 wooden skewers

## Procedure

1. Place one strawberry in each plastic bag. Seal the tops of the bags and knead the bags to completely mash the strawberries. Add 3 pinches table salt to each bag and mix again.

## Camper Notebook

Your label may differ.

2. Your leader will tell you which cleaner solution to make. Prepare the cleaner solution by putting 2 teaspoons (10 mL) water into a small cup, then doing one of the following:
   - For a soap solution, use a laboratory spatula or butter knife to scrape less than ¼ teaspoon (1.25 mL) Ivory bar soap into the cup. Stir until the soap is dissolved.
   - For a dishwashing detergent solution, add 5 drops dish detergent into the cup and swirl.
   - For a shampoo solution, add 10 drops shampoo into the cup and swirl.

3. Add all of the cleaner solution you made in step 2 into one bag containing a mashed strawberry and label the bag with the type of cleaner solution you prepared. Add 2 teaspoons (10 mL) water to the other bag containing a mashed strawberry and label the bag "no cleaner." (See left.) Gently knead each bag for about 2 minutes.

4. Put aside the bags of strawberries for a moment. Label two test tubes like the bags in step 3 (one with the type of cleaner solution you prepared and one with "no cleaner"). Add ice-cold isopropyl alcohol provided by your leader to each test tube to a height of about 2 inches. Store the test tubes in an ice bath. (See right.) These test tubes will be used to store the wooden skewers at the end of step 7.

Your label may differ.

*Setup described in step 4 for storage of DNA collected in steps 7 and 8*

5. Label two clean small cups like the bags in step 3. Pour each strawberry mixture through a funnel lined with filter paper and into the appropriate cup. (See left.) Be sure to clean the funnel and use new paper after filtering the first mixture. Discard the bags and filters containing the strawberry solids into the trash.

6. Label two clean test tubes like the bags in step 3. Pour the strawberry cell extracts from the cups into the corresponding test tubes to a height of about 1 inch. Place the test tubes in the ice bath.

## Camper Notebook

**Tip...**
It may be necessary to return the test tube to the ice bath periodically during step 7. Since DNA is less soluble in cold solutions, you will be able to collect more DNA when the test tube is kept cold.

**7.** Carefully pour about an inch of ice-cold isopropyl alcohol provided by your leader on top of the cold, cleaner-treated extract. Do not mix or swirl. Observe the interface between the strawberry extract and isopropyl alcohol. If milky strands (DNA) form at the interface, use a twisting motion to wind the milky fibers onto a wooden skewer. Continue twisting until no more DNA can be wound on the wooden skewer. Store the skewer in the appropriately labeled test tube you prepared in step 4 that contains just isopropyl alcohol. Keep the test tube in the ice bath.

**8.** Repeat step 7 with the cold strawberry extract not treated with cleaner.

**9.** Briefly remove both skewers from the isopropyl alcohol and compare the results of the two extractions. Return the skewers to the test tubes to save them for the Wrap-Up.

**Q1:** How do the two skewers differ? Explain the results.

**Q2:** Strawberry DNA is genetic material located inside the cell walls of strawberry cells. In step 8, how much DNA did you isolate from the cold strawberry extract not treated with cleaner? Why?

**10.** To clean up, throw away the skewers and DNA in the trash, rinse the strawberry extracts and other liquids down the drain, and rinse out the equipment.

# Colorful Lather Printing

## Overview

In this activity, you will use shaving cream, a common soap lather, to create beautiful colored patterns. At the same time you'll explore the chemistry of soap.

## Materials

- ✓ newspaper
- ✓ food colors in dropper bottles
- ✓ 3–4 pieces of a nonglossy, sturdy paper
  ☞ *Index cards, copy paper, or art paper work well.*
- ✓ small clear cup
- ✓ water
- ✓ aerosol shaving cream (standard white type)
- ✓ paper plate
- ✓ cooking spatula or craft stick
- ✓ toothpicks
- ✓ dropper or straw
- ✓ paper towels for cleanup

## Procedure

1. Cover your work surface with newspaper. (Food color can stain wood surfaces.)

2. Place one drop of food color onto a clean sheet of paper. Notice the degree to which the color spreads on the paper.

3. Fill a cup about half-full with room-temperature water. Without stirring, add one drop of food color to the water. Notice the degree to which the color spreads through the water.

4. Dispense a pile of shaving cream roughly the size of a your fist onto the paper plate. Using a spatula or craft stick, shape the shaving cream so that the top surface is nearly flat and the area of this top surface is slightly larger than the paper that you will be marbling. Apply drops of several different colors of food color to different locations on the shaving cream. (A total of 6–8 drops works well.) Notice the degree to which the color spreads through the shaving cream.

# Camper Notebook

5. Drag a toothpick through the colored drops on the shaving cream to create patterns with the color. Drag colored shaving cream into uncolored areas or uncolored shaving cream into colored areas. Try making different patterns. Straight lines, curved lines, parallel lines, and spirals will all produce different effects.

6. Press the paper onto the surface of the shaving cream. You will notice that the paper becomes wetted by the colored shaving cream, and some of the color pattern may show through the paper. Pull the paper off the shaving cream. Some shaving cream will adhere to the paper. Scrape off the excess shaving cream close to the paper using the spatula (or craft stick) and return the excess shaving cream to the original pile. The patterns of color that you created on the surface of the shaving cream should transfer to the paper.

7. You can make additional marbled papers by repeating steps 4–6 if you wish. Otherwise, move on to step 8.

8. Using the spatula (or craft stick), mix the pile of colored shaving cream until it is one uniform color. If the color is very pale, mix in a few more drops of food color.

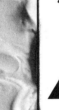

9. Apply a single drop of water to the surface of the colored shaving cream and observe what happens. You may wish to apply additional drops of water at different places on the surface for design purposes. Now try repeating steps 5–6 with the shaving cream mixture that remains.

⚠ *Even though shaving cream is formulated to be safe on skin, it can become irritating if left on your skin for too long. Be sure to wash your hands when you are finished with this activity.*

**Q1:** Describe the spreading you observed when dropping food color onto clean paper (step 2), into water (step 3), and onto shaving cream (step 4).

**Q2:** Shaving cream is a lather, similar to a foam. A foam is a colloid (a gas trapped within a liquid). Do research to answer these questions: What other common products are foam or lather colloids? How are colloids, in general, different from solutions? What do solutions and colloids have in common?

# Surfactants

## Overview

Why do soaps and detergents belong to a group of chemicals called surfactants? Experiment with soapy water to find out.

## Materials

- ✓ 2 new plastic cups (free from any soap or detergent residue) such as 3-ounce (90-mL) bathroom cups or 9-ounce (270-mL) tumblers
- ✓ same-sized counters that sink in water (such as pennies or plastic craft beads)
- ✓ distilled water, deionized water, or purified drinking water without added minerals
- ✓ soapy water (prepared by leader)
- ✓ dropper or disposable pipet
- ✓ clean surface (such as waxed paper or bottom of plastic cup)
- ✓ 2 pennies
- ✓ paper towels for spills and clean up

## Procedure

**Test with Cups of Water**

1. Fill one new, clean cup to the very top with purified water. In the data table at the end of this activity, sketch a side view of the cup to show what the top surface of the water looks like before adding the counters in step 2.

2. Gently add counters to the cup one at a time without splashing the water. In the data table, record the actual number of counters you added just before the water in the cup spills over the side. Sketch a side view of the cup to show the shape of the surface of the water after adding the counters and just before the water spills over the side.

**Q1:** What happens to the shape of the surface of the water?

3. Fill another cup to the very top with soapy water. In the data table, record the name and concentration of the detergent solution. Sketch the top surface of the water.

## Camper Notebook

4. Repeat step 2 with the soapy water cup.

**Q2:** What happens to the shape of the surface of the soapy water?

**Q3:** Do you think the number of counters that can fit in the cup containing soapy water depends on the type and brand of detergent and the amount of detergent used to make the soapy water? How can you find out?

### Test with Pennies

5. Fill a clean dropper or disposable pipet with purified water. Drop one drop on a clean surface such as a piece of waxed paper or the bottom of an upside down plastic cup. After observing the size of one drop, predict how many drops of purified water you think will fit on the heads side of a penny. Write your prediction in the data table at the end of this activity.

6. Counting as you drop, use the dropper to carefully place drops of purified water onto the heads side of a penny. In the data table under trial 1, record the actual number of drops added just before water spills off the penny. Dry the penny and repeat for two more trials. Calculate an average.

7. Place a second penny heads up. Fill a dropper with soapy water. In the data table, record the name and concentration of the detergent used and your prediction of how many drops of soapy water will fit on the penny. Then, counting as you drop, carefully place drops of soapy water on the heads side of the penny. In the data table under trial 1, record the actual number of drops added just before the water spills off the penny. Dry the penny and repeat for two more trials. Calculate an average.

**Q4:** Describe what is different about the shape of the liquid surfaces on the two pennies.

**Q5:** Have you observed the high surface tension of water in other instances before today? Where?

Disease Control Is in Your Hands—Surfactants

## Camper Notebook

| | Test with Cups of Water ||
|---|---|---|
| | Sketch Top Surface Before Adding Counters | Sketch Top Surface Just Before Water Spills Over |
| cup of purified water | | number of counters _____ |
| cup of soapy water containing _____ detergent name ( _____ ) detergent concentration | | number of counters _____ |

| | Test with Pennies ||
|---|---|---|
| | Predicted Number of Drops | Actual Number of Drops |
| penny holding purified water | | trial 1 |
| | | trial 2 |
| | | trial 3 |
| | | average |
| penny holding soapy water containing _____ detergent name ( _____ ) detergent concentration | | trial 1 |
| | | trial 2 |
| | | trial 3 |
| | | average |

# Blowing Bubbles

*Blow a large bubble*

*Measure the wet ring*

## Overview

Soap bubbles make beautiful containers for gases. Admire these vessels and discover which dishwashing detergent forms the biggest soap bubbles.

## Materials

- ✓ 3 different kinds of dishwashing detergent solutions in plastic cups (prepared by leader)
- ✓ teaspoon or 5-mL graduated cylinder
- ✓ 3 large trays (such as cookie sheets lined with black plastic or large dark-colored Styrofoam® meat trays)
- ✓ 3 plastic coffee stirrer straws or drinking straws
- ✓ metric ruler or meterstick
- ✓ paper towels
- ✓ vinegar in spray bottle
- ✓ (optional) spot markers such as paper clips

## Procedure

1. Pour 1 teaspoon (5 mL) of the first bubble solution into the center of a clean tray. Tip the tray to allow the bubble solution to completely cover the surface of the tray.

2. Dip one end of a coffee stirrer straw or drinking straw into the cup of bubble solution. Place the end of the straw so that it just touches the center of the soapy surface of the tray.

3. Gently and continually blow into the clean end of the straw. Try to blow one large bubble dome. Continue blowing until the bubble pops. The bubble will leave a wet ring on the tray that will be visible for at least a few seconds.

4. Use a metric ruler or meterstick to measure across the widest part of the ring left by the bubble. This is the diameter of the bubble. In the data table at the end of this activity, record the diameter of the bubble for trial 1 and the brand of the dishwashing detergent you used.

**Tip...** If the ring disappears faster than you can do step 4, use spot markers such as paper clips to mark each side of the widest part of the wet ring.

## Camper Notebook

5. Repeat steps 3 and 4 at least two more times with the same bubble solution. Calculate and record the average bubble diameter for the bubble solution you just tested.

**Q1:** Are the sizes of the bubbles blown approximately the same? Why or why not?

6. Repeat steps 1–5 with the remaining dishwashing detergents. Make sure to use a different tray and straw for each solution tested.

**Q2:** Why do you use a different tray and straw for each solution tested?

**Q3:** Although consumers often think bubbles (suds) and lather are important for cleaning ability, bubbles and lather are not necessary for cleaning. Give an example of a cleaning situation when suds are desirable.

### To clean up...
- Throw away all straws.
- Wipe up any excess bubble solution from the tabletop with dry paper towels.
- Spray the surface with vinegar to remove the soap film and wipe the table dry with fresh paper towels. (Do not use water because that would create more suds.)

|  | Bubble Diameter | | |
|---|---|---|---|
|  | Brand of Dishwashing Detergent: _____ | Brand of Dishwashing Detergent: _____ | Brand of Dishwashing Detergent: _____ |
| trial 1 |  |  |  |
| trial 2 |  |  |  |
| trial 3 |  |  |  |
| average |  |  |  |

# Watching Granny Smith Rot

*See how important it is to wash your hands before handling food.*

FYI: You will need three friends to do this activity.

## What You'll Need:

- apples, washed in advance
- small knife, cutting board, and peeler, washed in advance
- *Use knife and peeler with adult supervision.*
- access to a sink with soap, warm water, and paper towels
- 4 new pint- or quart-sized zipper-type freezer bags
- permanent marker to label bags
- regular soap (not antibacterial)
- antibacterial soap
- alcohol-based hand sanitizer

## What You'll Do:

>> 1. Label the four zipper-type freezer bags as follows: "unwashed hands," "regular soap," "antibacterial soap," and "alcohol-based hand sanitizer."

>> 2. Wash your hands thoroughly with regular soap and warm water, lathering the hands for at least 20 seconds. If you haven't done so recently, thoroughly wash the knife, cutting board, and peeler with soap and warm water. Carefully cut an apple in quarters on the cutting board. Peel one quarter of the apple (letting the peel fall onto a paper towel) and place the peeled piece of apple in the bag labeled "regular soap." Tightly close the bag. Dispose of the paper towel and peel. Wash the peeler with soap and warm water.

>> 3. Choose one friend who has not washed his or her hands in several hours to peel another quarter of the apple (over a paper towel). Have that person place the peeled piece of apple in the bag labeled "unwashed hands" and tightly close the bag. Dispose of the paper towel and peel. Wash the peeler with soap and warm water.

>> **4.** Ask a second friend to thoroughly wash his or her hands with antibacterial soap and then to peel one of the remaining apple quarters and place it into its bag. Close the bag and clean up as before including rewashing the peeler.

>> **5.** Ask a third friend to clean his or her hands with alcohol-based hand sanitizer as directed on the bottle and repeat the apple peeling, bagging, and cleanup procedures as before.

>> **6.** Place all of the bags in a warm place.

>> **7.** Without opening the bags, observe the apples once a day for at least one week. Record your observations in the data table at the end of this activity using both words and illustrations. You may want to take digital pictures of the bags each day to compare later.

> **Important...**
> For health reasons, do NOT open the bags! When you are finished with the experiment, throw the bags in the trash with the apples still inside.

## Questions to Consider:

- Review your observations. What generalizations can you make about all of the apple pieces regardless of how they were handled?
- What can you conclude about the effects the various hand-washing methods played in your final observations?

## What's the Deal?

The apple pieces will typically turn brown within 30 minutes to an hour after being cut. This is due to the reaction of the apple with oxygen in the air and is not a result of microbial activity. The appearance of bacterial or mold growth and the resulting rot will not be visible to the naked eye for several days. Typically, an apple cut with unwashed hands grows mold within about five days and is covered in mold after about seven days. An apple cut with washed hands only (regular or antibacterial soap) has minor traces of mold after a week. An apple cut after using hand sanitizer has no visible mold growth after seven days.

This activity vividly shows how unwashed hands transfer bacteria and mold spores to the apple pieces. Washed or sanitized hands transfer few or no microorganisms to the apple. Washing hands before preparing and eating food can help prevent the spread of illness.

**Tip...** You may want to take digital pictures of the bags each day to compare later. Do not open the bags when taking the pictures!

## Observations

|        | Regular Soap | Unwashed Hands | Antibacterial Soap | Alcohol-Based Hand Sanitizer |
|--------|--------------|----------------|--------------------|------------------------------|
| start  |              |                |                    |                              |
| Day 2  |              |                |                    |                              |
| Day 3  |              |                |                    |                              |
| Day 4  |              |                |                    |                              |
| Day 5  |              |                |                    |                              |
| Day 6  |              |                |                    |                              |
| Day 7  |              |                |                    |                              |
| Day 8  |              |                |                    |                              |
| Day 9  |              |                |                    |                              |
| Day 10 |              |                |                    |                              |

**Take-Home Activity: Watching Granny Smith Rot**

© 2007 Terrific Science Press™
The publisher takes no responsibility for use of any materials or methods described in this monograph, nor for the products thereof. This publication was made possible by Grant Number 1 R25 RR16301-01A1 from the National Center for Research Resources (NCRR), a component of the National Institutes of Health (NIH). Its contents are solely the responsibility of the authors and do not necessarily represent the official views of NCRR or NIH.

# How Clean Is Your Clan?

*How well do your friends and family wash their hands? If you ask them, there's a good chance they'll tell you they wash thoroughly. But do they?*

FYI: You will need friends and family to help with this activity.

## What You'll Do:

>> 1. Ask your friends and family to rank how well they routinely wash their hands according to the following scale:

1 = very poor
2 = poor
3 = fair
4 = good
5 = very good
6 = excellent

Record their names and self-rating responses in the following Data Table 1.

| How Clean Is Your Clan? Data Table 1 | | |
|---|---|---|
| Name | Self-Rating Response | Observation Score |
|  |  |  |
|  |  |  |
|  |  |  |
|  |  |  |
|  |  |  |
|  |  |  |
|  |  |  |
|  |  |  |
|  |  |  |

>> 2. Observe each person washing his or her hands. Ask them to try hard not to do anything differently just because you're watching. In Data Table 2 (below), score each person's hand-washing habits by giving one point for each of the following correct hand-washing techniques.*

- uses soap (1 point)
- does not hold hands under running water while lathering them (1 point)
- rubs hands together to lather soap for at least 15–20 seconds (1 point)
- washes all surfaces, including fronts and backs of hands up to the wrist, on and between all fingers, and under fingernails (1 point)
- rinses with running water (1 point)
- dries hands with a clean towel (1 point)

| How Clean Is Your Clan? Data Table 2 | | | | | | |
|---|---|---|---|---|---|---|
| | Observations | | | | | |
| Name | Uses Soap | Lathers Hands Away from Faucet | Lathers At Least 15–20 Seconds | Washes All Hand Surfaces | Rinses with Running Water | Dries with Clean Towel |
| | | | | | | |
| | | | | | | |
| | | | | | | |
| | | | | | | |
| | | | | | | |
| | | | | | | |
| | | | | | | |
| | | | | | | |
| | | | | | | |

*Criteria are based on the Clean Hands Campaign of the American Society for Microbiology. (See www.washup.org.)

>> 3. Total up the points and record the observation score in the last column of Data Table 1. Discuss the results with each friend and family member.

Take-Home Activity: How Clean Is Your Clan?

CAMPER

# Topic 2

## Water Purification

### List of Activities

- ✓ Back to Its Elements ............................................................. 56
  *Observe how water can be broken apart into its component elements.*

- ✓ Distinguishing Water Samples .......................................... 58
  *Investigate the differences between tap water and different bottled waters using super-absorbing crystals.*

- ✓ Osmosis with Eggs ............................................................. 61
  *Observe the process of osmosis using decalcified eggs. (Reverse osmosis is a common water treatment process.)*

- ✓ Water Taste Test ................................................................. 65
  *Do a water taste test with various brands of bottled water.*

- ✓ Investigating Mineral Content ......................................... 67
  *Determine the mineral content of water samples based on their acidity or basicity.*

- ✓ How Hard Is Your Water? .................................................. 69
  *Estimate the relative amounts of dissolved minerals in different types of water.*

- ✓ Purifying Surface Water .................................................... 71
  *Investigate the role of flocculation in purifying water.*

- ✓ How Much Iodine Is Present? ........................................... 74
  *Learn one method of monitoring chlorine levels in tap water. (Iodine is tested instead of chlorine because it is safer to handle.)*

- ✓ Removal of Iodine from Water ......................................... 76
  *Explore the effectiveness of activated charcoal (commonly used in tap water purification) to remove residual chlorine. (Iodine is tested instead of chlorine because it is safer to handle.)*

- ✓ Take-Home Activity: What's Really in the Bottle? ......... 78
  *Examine bottled water labels to learn the different types of bottled water.*

- ✓ Take-Home Activity: The Amazing Water Maze ............ 84
  *Take a journey through the water maze. What will you label the bottled water?*

# Back to Its Elements

## Overview

Observe how water can be broken apart into its component elements using an electric current.

## Materials

- ✓ 2 small Beral pipets (disposable polyethylene transfer pipets) or transparent straw
- ✓ scissors
- ✓ metric ruler
- ✓ 2, 10-inch (25-cm) long floral wires (prepared by leader)
- ✓ hot-melt glue gun and glue (if using straws)
- ☞ *Use the hot-melt glue gun with adult supervision.*
- ✓ electrolyte and indicator solution (prepared by leader)
- ✓ petri dish or other small shallow bowl
- ✓ 9-volt battery
- ✓ goggles
- ✓ (optional) wire to make a stand
- ✓ (optional) tape
- ✓ (optional) alligator clips
- ✓ (optional) 9-volt snap connector and alligator clips

## Procedure

1. Prepare the electrodes in one of two ways.

   - Cut off the long stems of two pipets so that only about ¼ inch (0.5 cm) of the stems remain as shown at left. For each pipet, follow steps a–c.

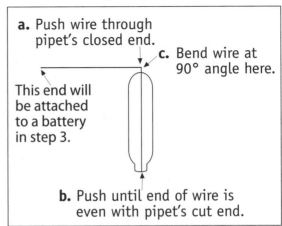

a. Push wire through pipet's closed end.

c. Bend wire at 90° angle here.

This end will be attached to a battery in step 3.

b. Push until end of wire is even with pipet's cut end.

# Camper Notebook

- As an alternative, cut a straw to make two 1½ inch (about 3–4 cm) lengths. For each straw, follow steps a–d.

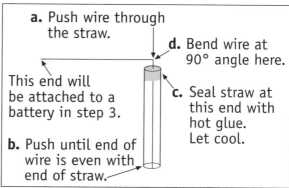

a. Push wire through the straw.

b. Push until end of wire is even with end of straw.

c. Seal straw at this end with hot glue. Let cool.

d. Bend wire at 90° angle here.

This end will be attached to a battery in step 3.

2. Fill each electrode with the electrolyte and indicator solution prepared by your leader. Cover the bottom of a container (petri dish or shallow bowl) with the same solution. A very shallow layer about ¼ inch (0.5 cm) deep is sufficient.

3. Quickly invert each filled electrode into the container's solution so the solution stays in the electrode and the electrode's long wire sticks out of the top as shown. Keep the electrodes upright at all times by either holding them, supporting them with a wire stand you make (Method A), or attaching them to the inside of the container with tape (Method B). Be sure the electrodes stay upright.

4. Connect one wire to each terminal of a 9-volt battery by one of the following methods:
   - Twist the wire around the terminal.
   - Twist the wire onto an alligator clip and clip the alligator clip to the terminal.
   - Twist the wire onto an alligator clip. Place a 9-volt snap connector on the battery. Clip the alligator clip to the wire coming off the connector.

   Most importantly, the electrode wires must not make contact with each other.

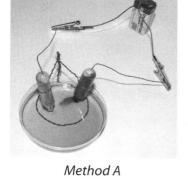

*Method A*

5. After the two wires are connected to the battery terminals, make observations for several minutes.

**Q1:** What changes are observed as the reaction proceeds?

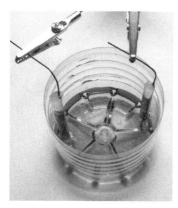

*Method B*

**Q2:** How do the relative amounts of gases produced at each electrode indicate which gas is formed?

Water Purification—Back to Its Elements

# Distinguishing Water Samples

## Overview

Water-absorbent polymers in certain products (such as disposable diapers and water-storing crystals for potting soil) may absorb liquid many times their volume. How much liquid will these polymer crystals absorb? Is the amount of water absorbed by these crystals different for different kinds of bottled water and other beverages?

## Materials

- ✓ 9-ounce (about 270-mL) or larger clear cups (one for each liquid sample)
- ✓ permanent marker or self-stick labels
- ✓ granular water-storing polymer product having crystals measuring about 2–4 mm in diameter
- ✓ tap water
- ✓ 2 or 3 different types of bottled water (such as distilled water, spring water, and mineral water)
- ✓ Gatorade® or similar sports drink
- ✓ 1-cup (250-mL) liquid measuring cup with metric markings
- ✓ strainer (such as tea strainer or plastic cup with pushpin holes in the bottom)

## Procedure

1. Label a cup for each sample to be tested. Be sure to include tap water. In the data table at the end of this activity, record the names of the liquids you will test. Read the bottle labels and record the ingredient list for each sample.

2. Place 10 polymer crystals that are about 2–4 mm wide into each of the labeled cups. Since the crystals are irregularly shaped and some are closer to 2 mm while others are closer to 4 mm, try for a similar sample of 10 for each cup. (In other words, do not initially select only the largest crystals, leaving smaller crystals for the other cups.)

# Camper Notebook

3. Add 150 mL of the appropriate liquid to each of the labeled cups. Allow the cups to sit several hours or overnight.

4. Hold the strainer over the measuring cup and pour the contents from one sample cup into the strainer. Once the liquid has drained into the measuring cup, return the crystals to their original (now empty) cup. (You are saving the crystals for later comparisons.) Read the volume of liquid you collected in the measuring cup in milliliters as accurately as possible. Record the volume in the data table. Rinse the liquid down the drain.

5. Calculate the volume of liquid absorbed by the crystals using the following equation:

volume of liquid initially added to the cup − volume of liquid collected after straining = volume of liquid absorbed by the crystals

Record your result in the data table.

6. Repeat steps 4 and 5 for each sample.

7. Examine the crystals that you returned to the cups in step 4. Are the crystals transparent (clear) or cloudy? Are air bubbles present? How do their swollen volumes compare? Record your observations in the data table.

8. In the data table, rank the typical swollen crystal size of each sample from smallest (number 1) to largest.

**Q1:** Look at the data that you collected. What (if any) trends do you observe with regard to the amount of liquid absorbed by the crystals and the ingredients/water sources listed for the samples?

**Q2:** Based on your experimental results and conclusions, which, if any, water samples have labels that are not accurate and/or clear? Describe the problems.

**Important...**
When you are done with the crystals, throw them in the trash. *Do not* dump the crystals down the drain because they can clog plumbing.

## Camper Notebook

| Sample | Ingredients and/or Water Source Listed on Label | Volume of Liquid | | | Rank |
| --- | --- | --- | --- | --- | --- |
| | | Liquid Added to Cup (step 3) | Liquid Collected After Straining (step 4) | Liquid Absorbed by Crystals (step 5) | |
| tap water | Don't write anything. | 150 mL | | | |
| | | observations: | | | |
| | | 150 mL | | | |
| | | observations: | | | |
| | | 150 mL | | | |
| | | observations: | | | |
| | | 150 mL | | | |
| | | observations: | | | |
| | | 150 mL | | | |
| | | observations: | | | |
| | | 150 mL | | | |
| | | observations: | | | |

# Osmosis with Eggs

## Overview

Most water must be treated before it is safe to drink. Reverse osmosis is one method for purifying water. What is osmosis and reverse osmosis? How do these processes work? In this activity, you'll use eggs to observe the effects of osmosis, and you'll research several other ways to purify water.

## Materials

- ✓ 2 large, uncracked hard-boiled eggs
- ✓ one or more of the following to measure the eggs:
  - balance capable of measuring to 0.1 g
  - egg template (provided at the end this activity) copied on heavy paper or light cardboard
  - graduated cylinder at least 5.6 cm wide *or* cup large enough to totally submerge an egg in, a pan to set the cup into, and a teaspoon or other volume measure
- ✓ 2 cups or wide-mouthed containers large enough to totally submerge an egg in
- ✓ water
  - ☞ *Tap water works okay, but bottled purified drinking water works better in step 3.*
- ✓ light corn syrup
- ✓ paper towels
- ✓ (optional) scissors

## Procedure

1. Observe closely as your leader puts an egg into vinegar.

**Q1:** What evidence do you see that the shell is reacting with the vinegar?


2. Since the reaction of the eggshell and vinegar is slow, your leader has prepared the eggs you will use. Measure each egg by any or all of the following methods and record the results in the "Start" column in the data table at the end of this activity.

Water Purification—Osmosis with Eggs

## Camper Notebook

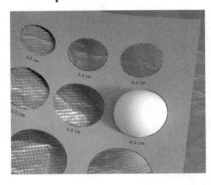

**Method 1:** Weigh the egg carefully on a balance as accurately as possible. (This is the best method to use when you are going to remeasure after only 1½ hours.)

**Method 2:** Determine the approximate diameter across the short axis (width) of the egg. If the leader hasn't already done so, cut out the circles on the egg template. Select the smallest template hole that the egg will slip through on its own. Carefully put the egg through the hole using the same end of the egg each time (either the pointed end or the wide end). Have a hand underneath the template to catch the egg.

**Method 3:** Measure the volume of the egg by determining the amount of water it displaces. Do one of the following.

- Add water to a graduated cylinder about halfway up and record the volume. Add the egg and, making sure the egg is completely covered with water, record the second volume. The volume of the egg is the second volume reading minus the first volume reading.

- If you do not have a graduated cylinder, place a cup into a pan and fill the cup to the brim with water. Carefully add the egg to the cup, making sure that the egg is completely covered with water. The egg will displace a volume of water equal to its volume, and this water will overflow into the pan. Carefully remove the cup containing the egg without spilling any additional water into the pan. Use a teaspoon or other volume measure to determine the volume of water that spilled into the pan. This volume equals the volume of the egg.

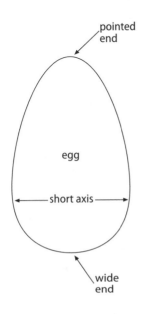

3. Label one cup "water" and the other "light corn syrup." Place one egg in each cup. Completely cover the eggs with the designated liquids.

**Q2:** What do you predict will happen to the sizes of the two eggs over time? Write down your prediction.

# Camper Notebook

4. After 24, 48, 72, and 96 hours, dry the eggs by gently patting them with paper towel. Measure the eggs using the same method you used in step 2 and record the results in the data table. If you can only check your eggs the same day, let the eggs soak in the cups (step 3) for 1½ hours before drying and measuring them using Method 1.

5. To help illustrate the results, graph the data for both eggs. (See the sample graph setup at left.)

6. After your leader has given you an explanation of osmosis, answer the following questions.

**Q3:** Which egg became larger in the process? Why do you think this happened?

**Q4:** In what direction does water move through the egg membranes when the egg is placed in the corn syrup? What's your evidence for this?

**Q5:** Research and list ways to purify water. Which methods involve a physical change and which involve a chemical change?

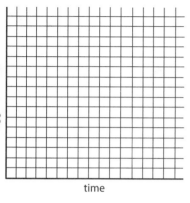

*Sample graph setup*

| Egg Measurements | | | | | | |
|---|---|---|---|---|---|---|
| | **Start** | **1½ Hours (optional)** | **24 Hours (1 day)** | **48 Hours (2 days)** | **72 Hours (3 days)** | **96 Hours (4 days)** |
| egg in water | | | | | | |
| egg in light corn syrup | | | | | | |

Water Purification—Osmosis with Eggs

# Egg Template

Copy this onto heavy paper or light cardboard and then cut out the circles. Make the cut edges as smooth as possible.

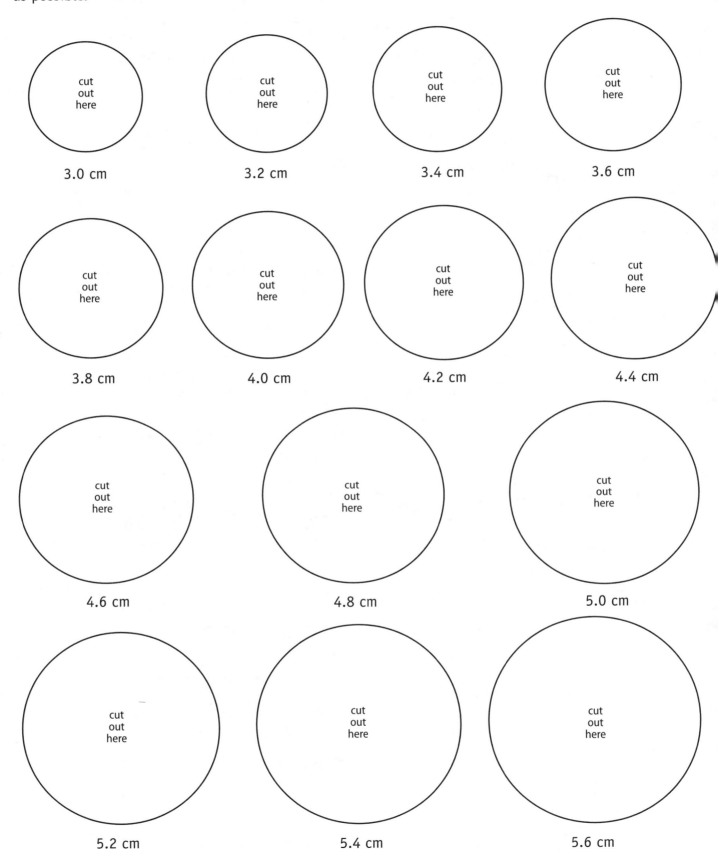

# Water Taste Test

## Overview

Do you like tap water or a specific bottled water? In this activity, you will taste unidentified drinking water samples, rank them according to taste, and then attempt to identify them based on their appearance and taste alone.

## Materials

- ✓ cups of water samples (provided by leader)
- ✓ room where eating and drinking is permitted
- ✓ (optional) unsalted crackers to cleanse your palate between tastings

## Procedure

1. Look closely at each water sample. In the data table at the end of this activity, use a few words or phrases to describe the appearance of each sample.

2. Smell each water sample. In the data table, use a few words or phrases to describe the odor.

3. Sip each water sample. In the data table, use a few words or phrases to describe the taste. It is okay to retaste the samples. You may want to eat a piece of unsalted cracker between tasting samples to cleanse your palate.

4. Use your own opinions regarding taste to rank the water samples from best tasting to worst tasting, with 1 being the best. Record the ranks in the data table. Retaste each sample if necessary.

5. From the general list of water brands (or sources) provided by the leader, guess the identity of each water sample and enter your guesses in the data table.

6. From the leader, find out the actual water sample identities and prices per unit volume. Fill in the last two columns of the data table.

## Camper Notebook

**Q1:** What, if any, results of the taste test are surprising?

**Q2:** What factors may influence the taste of the different waters?

| Water Sample | Appearance (step 1) | Odor (step 2) | Taste (step 3) | Rank (step 4) | Guess the Water Sample (step 5) | Water Sample Identity (step 6) | Price per Unit Volume (step 6) |
|---|---|---|---|---|---|---|---|
| A | | | | | | | |
| B | | | | | | | |
| C | | | | | | | |
| D | | | | | | | |
| E | | | | | | | |
| F | | | | | | | |

# Investigating Mineral Content

## Overview

In this activity, you determine the relative mineral content of different water samples based on their acidity or basicity. Most water samples are nearly neutral; this means that the water is neither very acidic nor very basic. If a water sample is basic, this indicates the presence of minerals. A simple way to test the acidity or basicity of water is to use turmeric (a spice), which contains an indicator that is yellow if the sample is neutral or acidic and red if the sample is basic. In this activity, you will use turmeric-treated paper to test different water samples.

## Materials

- ✓ manila file folder or other paper that is *not* acid-free
- ✓ turmeric solution (prepared by leader)
- ✓ paint brush, foam brush, or sponge
- ✓ distilled water in a container
- ✓ tap water in a container
- ✓ several different bottled waters
- ✓ saturated baking soda ($NaHCO_3$) solution (prepared by leader)
- ✓ vinegar (5% acetic acid)
- ✓ transfer tools such as droppers, disposable pipets, drinking straws, or cotton swabs (one for each water sample to be tested)
- ✓ (optional) hair dryer

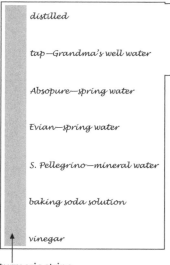

*Example of setup*

## Procedure

1. Dip a brush or sponge in turmeric solution and paint a long stripe on the left side of your paper. (See example at left.) Allow the solution to dry.

2. Write the name of each sample to be tested directly on the paper to the right of the turmeric stripe, separating the names along the stripe as much as you can. (See example at left.) Be sure to include the saturated baking soda solution and the vinegar as samples. Leave room to write additional information near each sample name.

Water Purification—Investigating Mineral Content

## Camper Notebook

3. For each bottled water sample, read the label or search the Internet to try to determine the source of the water (such as artesian, spring, mineral, sparkling, municipal, well, or purified) or its mineral content. (Some labels and Internet sites won't reveal this information.) Write down what you are able to find out next to the bottle's brand name. Also, write down where the tap water came from.

4. Using a dropper or other transfer tool, place 1 drop of each water sample on the turmeric stripe next to its name. To avoid cross-contamination, be sure to use a different transfer tool for each water sample.

5. Allow the water drops to dry. Air drying usually takes about 20 minutes. You can speed up the process by drying the drops with a hair dryer.

6. Record the color of the dried water drop next to the name of each water sample. Try to rank the samples from highest to lowest mineral content.

**Q1:** How many of the water samples you tested have high mineral content? Which ones are they? How did you determine this?

**Q2:** Look for patterns between the color of the dried drops and the water sources. Do any of the results appear to contradict what the water label says about either the source or mineral content?

**Q3:** How do the tap water results compare to spring and mineral water results?

> **Remember...**
> - The sample is neutral or acidic if its dried drop is yellow.
> - The sample is basic if its dried drop is red.
> - The darker red the dried drop, the more minerals the sample contains.

# How Hard Is Your Water?

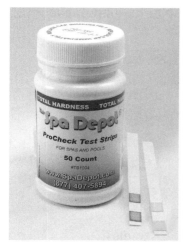

## Overview

As rainwater passes through soil and underground rocks, it dissolves some minerals (such as calcium and magnesium ions) and other substances. Many bottled water companies fill their bottles with this groundwater by taking the water from wells or springs. Minerals usually remain dissolved when the water is bottled. Water that contains a large amount of dissolved minerals is called "hard" water. Some people prefer to drink water that has a higher mineral content.

This activity allows you to determine the relative amounts of calcium and magnesium ions in different types of bottled water using water hardness test strips. These strips are calibrated to measure water hardness. Aquarium, swimming pool, and spa owners test their water with these strips to determine when water conditioning is needed.

## Materials

- ✓ small disposable cup for each water sample you are testing
- ✓ water hardness test strip for each water sample you are testing
- ✓ distilled water
- ✓ tap water
- ✓ several different brands of bottled water
- ☞ *Be sure to include at least one spring water and one mineral water.*

## Procedure

1. Based on what you have learned about water sources and dissolved minerals, predict whether each water sample is soft or hard. List the water samples and write your hardness prediction in the data table at the end of this activity.

2. To test if your prediction is correct, pour a small amount of the first water sample into a clean disposable cup.

## Camper Notebook

**Be safe...**
Do not drink the water after the strip is dipped into it.

**FYI...**
- In this case, parts per million (ppm) means one part calcium carbonate in one million parts of water.
- 1 ppm of hardness means one milligram of calcium carbonate per liter of water. (A liter of water is approximately equal to a kilogram of water.)
- 1 gpg is equal to 17.1 ppm.

3. Dip a hardness test strip into the water according to the directions on the test strip container. After removing the strip from the water, compare it to the color chart included in the test strip package. The colors on the chart may represent grains per gallon (gpg), parts per million (ppm), or level of water hardness. Record your results (including units) in the data table. Here's how the different measuring units compare:

| Grains per Gallon (gpg) | Parts per Million (ppm) | Description of Hardness |
|---|---|---|
| 0 gpg | 0 ppm | soft |
| 5 gpg | about 85 ppm | moderately hard |
| 10 gpg | about 170 ppm | hard |
| 20 gpg | about 340 ppm | extremely hard |
| 30 gpg | about 510 ppm | extremely hard |
| 50 gpg | about 850 ppm | extremely hard |

4. Repeat steps 2 and 3 for the remaining water samples.

**Q1:** Which of the tested waters is hardest?...least hard? How do the results compare to what you predicted (in step 1)?

**Q2:** What is the hardness value for the tap water you tested? Is your tap water hard or soft? How does your tap water result compare to the bottled waters you tested?

**Q3:** If you tested multiple brands of spring water, what is the range of hardness values you found? Discuss possible reasons for this range.

| Sample | Hardness Prediction | Actual Hardness |
|---|---|---|
| distilled | | |
| tap | | |
| | | |
| | | |
| | | |
| | | |

Water Purification—How Hard Is Your Water?

# Purifying Surface Water

## Overview

If your tap water comes from surface water such as a river or lake, you may wonder how the water is purified. Surface water sometimes looks dirty, cloudy, and yellow to brown in color. This activity shows you one way that surface water is treated to improve its quality.

## Materials

- ✓ soil
- ✓ paper to make a funnel
- ✓ 1 tablespoon and ⅛ teaspoon or other volume measures
- ✓ empty water bottle with a cap
- ✓ self-stick labels or permanent marker
- ✓ tap water
- ✓ liquid measuring cup
- ✓ 8 transparent plastic cups that hold at least 9–10 ounces (about 270 mL)
- ✓ alum (aluminum potassium sulfate dodecahydrate, $KAl(SO_4)_2 \cdot 12H_2O$)
- ✓ plastic spoons for stirring
- ✓ baking soda (sodium bicarbonate, $NaHCO_3$)
- ✓ funnel (such as a funnel cut from a 2-L plastic soft-drink bottle)
- ✓ 3 coffee filters
- ✓ (optional) household ammonia ($NH_3$ in water) and safety goggles

⚠ Be sure to wear safety goggles when using ammonia because it can damage eye tissue.

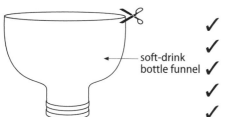

soft-drink bottle funnel

## Procedure

⚠ Do not drink any water samples in this activity. Samples that look clean may contain harmful bacterial that could make you sick.

1. Use a paper funnel to place 2 tablespoons (30 mL) of soil in an empty water bottle. Label the bottle "surface water." Add tap water to fill the bottle, put on the cap, and shake to mix. This mixture represents dirty surface water. It is okay if stones and heavy soil particles sink to the bottom of the bottle.

## Camper Notebook

2. Label cups as shown in the data table at the end of this activity.
3. Fill the cups as listed in the data table.
4. Add alum as listed in the data table. Stir until all of the alum dissolves. If particles in the surface water prevent you from observing when all of the alum has dissolved, stir the mixture as long as you stirred the tap water mixture.
5. Add baking soda as listed in the data table. As an alternative, you can put on safety goggles and add ⅛ teaspoon (½ mL) household ammonia instead of the baking soda. Be sure to wait 3–5 minutes before continuing with step 6.

**Q1:** Describe what happens in both cups when you add the baking soda.

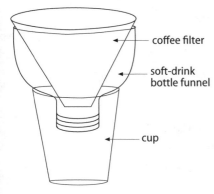

6. Label cups and filter the mixtures as listed in the data table. Each time you filter, place a clean coffee filter in the soft-drink bottle funnel and place the funnel on top of a new cup. Allow the mixture to drain into the cup.
7. Compare the physical characteristics of each water sample. Record your observations in the data table.

**Q2:** What is the result of adding aluminum hydroxide (alum and baking soda) and filtering the dirty surface water?

# Camper Notebook

## Purifying Surface Water Data Table

| Label Cups (step 2) | Fill Cups (step 3) | Alum (step 4) | Baking Soda (step 5) | Filter (step 6) | Observations (step 7) |
|---|---|---|---|---|---|
| tap water (control) | Add ½ cup (125 mL) tap water. | Don't do anything. | Don't do anything. | Don't do anything. | |
| tap water (treatment sample) | Add ½ cup (125 mL) tap water. | Add ⅛ teaspoon (about ½ mL) alum. Stir until dissolved. | Add ⅛ teaspoon (about ½ mL) baking soda. Stir well. Wait 3–5 minutes. | Label a new cup "tap water—treated and filtered." Filter into the new cup. | |
| surface water (control) | Add ½ cup (125 mL) surface water. | Don't do anything. | Don't do anything. | Don't do anything. | |
| surface water (treatment sample 1) | Add ½ cup (125 mL) surface water. | Add ⅛ teaspoon (about ½ mL) alum. Stir until dissolved. | Add ⅛ teaspoon (about ½ mL) baking soda. Stir well. Wait 3–5 minutes. | Label a new cup "surface water—treated and filtered." Filter into the new cup. | |
| surface water (treatment sample 2) | Add ½ cup (125 mL) surface water. | Don't do anything. | Don't do anything. | Label a new cup "surface water—filtered only." Filter into the new cup. | |

Water Purification—Purifying Surface Water

# How Much Iodine Is Present?

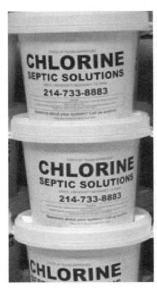

## Overview

To kill harmful bacteria in drinking water, municipal water companies often add chlorine. Chlorine is very effective at killing bacteria; however, it is also toxic at high levels to plants and animals. How do water companies keep track of the amount of chlorine they add to water so people don't get sick? This activity illustrates one method for monitoring chlorine levels. However, you'll be using iodine (another microbe-killing element), because it is easier to handle.

## Materials

- ✓ tap water
- ✓ graduated cylinder capable of accurately measuring 20.0 mL
- ✓ 3, 200-mL beakers or 3 clear plastic cups (at least 10-ounce size)
- ✓ white paper for a background
- ✓ water sample prepared by leader
  ⚠ *Do not drink the water sample because it contains iodine from a topical antiseptic.*
- ✓ disposable pipet, dropper, or small plastic squeeze bottle with dropper dispenser
- ✓ vitamin C solution prepared by leader

## Procedure

1. Measure 20.0 mL tap water and pour it into a beaker or cup. This is the clear control. Place the container on a white sheet of paper to use for the color comparison in step 3.

2. Measure 20.0 mL of your assigned water sample and pour it into another beaker or cup.

3. For trial 1, add one drop of the vitamin C solution to your water sample. Swirl to mix well. Compare your water sample to the clear control. If the water sample is not completely decolorized after mixing, add another drop of the vitamin C solution and mix well. Count the number of drops necessary to just decolorize the water sample. (Complete decolorization occurs when all of the iodine in the sample reacts with the vitamin C.)

**Important...**
Be sure to thoroughly mix each drop of vitamin C solution into the water sample before deciding to add another drop.

# Camper Notebook

4. In the data table at the end of this activity, record the number of vitamin C drops required to decolorize your water sample.

5. Repeat steps 2–4 two more times to get trial 2 and trial 3 results. Average your results (to obtain a more accurate number of vitamin C drops required to decolorize your water sample) and record your answer in the data table.

6. Use the average number of drops of vitamin C solution to calculate the approximate amount of iodine particles in your water sample. Here's why and how.
   - The decolorization of iodine occurs with one particle of vitamin C reacting with one particle of iodine. A low number of drops indicates little iodine is present, while a high number of drops means more iodine is present.
   - Finding the approximate number of iodine particles in your water sample requires some adjustment. Fill in the data table based on the following information.

   **By the way...**
   $1 \times 10^{18}$ is the same as 1 billion billion or 1 billion$^2$. That's 1 followed by 18 zeros!

   approximate iodine particles present = ☐* drops of vitamin C × $10^{18}$

   *Use the average value rather than one trial's value.

   If you are curious, your leader can explain where this adjustment factor comes from.
   - In a similar way, water companies can monitor chlorine by measuring how many drops of a solution are needed to react with all the chlorine in a water sample.

**Q1:** What is the approximate number of iodine particles in your sample?

**FYI...**
You will get data for the other water samples from other campers after everyone is done.

| Identity of Water Sample | Number of Vitamin C Drops | | | | Approximate Iodine Particles Present |
|---|---|---|---|---|---|
| | Trial 1 | Trial 2 | Trial 3 | Average | |
| A | | | | | × $10^{18}$ |
| B | | | | | × $10^{18}$ |
| C | | | | | × $10^{18}$ |
| D | | | | | × $10^{18}$ |

# Removal of Iodine from Water

## Overview

Some people don't want chlorine in their drinking water so they use home filtration systems that promise to remove some or all of the residual chlorine. How do home water filtration systems work? How effective are they at removing chlorine and other substances?

## Materials

- ✓ water sample prepared by leader
- ⚠ *The water sample contains iodine from a topical antiseptic. Do not drink the water sample, either before or after filtering.*
- ✓ 3 clear plastic cups (at least 10-ounce size) or 3, 200-mL beakers
- ✓ measuring cup, 100 mL graduated cylinder, or graduated beaker
- ✓ 2 self-stick labels
- ✓ white paper for a background
- ✓ 20 g (heaping 1 tablespoon) aquarium charcoal
- ✓ balance or tablespoon
- ✓ spoon or stirring rod
- ✓ filter (such as filter paper or coffee filter) and funnel (such as 60° laboratory funnel or funnel cut from a 2-L plastic soft-drink bottle)
- ✓ disposable pipet, dropper, or small plastic squeeze bottle with dropper dispenser
- ✓ vitamin C solution prepared by leader for "How Much Iodine Is Present?"
- ✓ (optional) pitcher water filter system

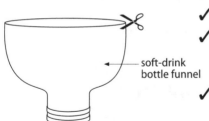

soft-drink bottle funnel

## Procedure

1. Measure ⅓ cup (or 75 mL) of your assigned water sample and pour it into a beaker or cup. Label this container "control." Place the container on a white sheet of paper. This sample will be used for the color comparison in step 4.

2. Measure another ⅓ cup (or 75 mL) of your water sample and pour it into another beaker or cup. Add 20 g (or a heaping tablespoon) of aquarium charcoal to the container. Stir the mixture for about 1 minute.

# Camper Notebook

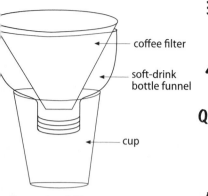

*Example of filter apparatus*

3. Pour the mixture into a filter apparatus to remove the solid charcoal, allowing the liquid to drain into a clean container. Label this container "filtered."

4. Compare the color intensity in the control with the color intensity in the solution treated with charcoal and filtered.

**Q1:** What evidence is there that the charcoal treatment removes some iodine?

5. Use the procedure outlined in the activity called "How Much Iodine Is Present?" to measure how much iodine is in the initial water sample ("control") and how much is left in the water sample after treating it with activated charcoal ("filtered.")

6. In the data table below, enter your results and those of other campers who began with water samples having different initial concentrations of iodine ($I_2$).

**Q2:** At what level of iodine in water does the charcoal filter remove all of the detectable iodine?

**Q3:** Based on your observations, what (if any) problems does activated charcoal filtration cause?

| Water Sample | Particles of $I_2$ Before Filtering | Particles of $I_2$ After Filtering |
|---|---|---|
| A | | |
| B | | |
| C | | |
| D | | |

*Water Purification—Removal of Iodine from Water*

# What's Really in the Bottle?

Millions of Americans drink bottled water every day. What's inside the bottle that makes it so special? If you know what to look for, the label on bottled water can tell you a lot about the product.

In this activity, you'll examine labels for fictional bottled water brands to identify the type of water in the bottles, how the water was treated, and where the water came from (its source). You may also want to examine the labels on actual bottled water. Perhaps you'll discover some interesting facts about your own favorite bottled water. An answer key is provided at the end of the activity, but please don't look at it until you've completed your investigation.

## What You'll Need:

- fact sheet (provided in this activity)
- data table (provided in this activity)
- label exhibits (provided in this activity)
- (optional) water bottles you have

## What You'll Do:

>> 1. Look over the fact sheet. You may want to refer to it as you complete the activity.

>> 2. Look at the label exhibits. For each label, record the requested information in the data table. For any information that is not provided on the bottle label, put a question mark (?) in the appropriate spot in the data table. If you can infer the information based on other parts of the label, put that information in brackets.

>> 3. After filling out the data table, look at the answer key. Does any of the information in the answer key surprise you?

### Questions to consider...

- What kinds of words and pictures do you see on the labels that would probably make consumers feel good about buying bottled water?
- Which sounds more appealing: "Tap Water in a Bottle" or "Purified Drinking Water?"

# What's Really in the Bottle? Fact Sheet

Bottled water is a term that covers a variety of water products sold in the United States. The U.S. Food and Drug Administration (FDA) regulates bottled water and has specific requirements for content and labeling. Not all water products fit the definition of bottled water. Enhanced water products such as Propel® and Fruit$_2$0® are considered beverages and are regulated differently by the FDA. These products may contain nutrients such as added vitamins, minerals, or fiber.

Nearly all drinking water in the United States, both bottled and tap, comes from either surface sources (lakes and rivers) or underground sources called aquifers. Underground water (also called groundwater) usually has more dissolved minerals than surface water. The following table lists several bottled water types that the FDA recognizes.

| Bottled Water Type | Explanation |
|---|---|
| Artesian Water | Artesian water is drawn from an underground source (aquifer) that is under pressure. Artesian water is tapped from the aquifer through wells or boreholes. In some artesian wells (called flowing artesian wells), the water is under enough pressure that it rises to the surface without the need of a pump. However, a bottled water can still be called artesian if the water is pumped and does not naturally rise to the surface. |
| Well Water | Well water comes from a borehole or well that taps the water of an aquifer. |
| Spring Water | Spring water has an underground source that naturally flows to the surface of the earth. Bottlers may collect the water directly from the spring or from a borehole near the location of the spring. Some brands of spring water undergo very little treatment due to the natural purity of the water. Other types of spring water may be treated with ozone or ultraviolet light to kill microorganisms. Many brands use micron filtration to remove particles. |
| Mineral Water | Mineral water naturally has at least 250 parts per million dissolved solids (minerals). This water must come from a protected underground source that is tapped from boreholes (wells) or at springs. After treatment to remove impurities, the water must have the same mineral content as it did when it came out of the ground. No additional minerals can be added. |
| Sparkling Water | Sparkling water comes from a well or spring and naturally contains carbon dioxide gas that gives it a "fizz" associated with seltzer or soda. After treatment to remove impurities, additional carbon dioxide can be added so that the level is the same as it was when it came from its source. |
| Drinking Water | Drinking water is another name for bottled water. All bottled water must be fit for human consumption (containing less than the allowed limits for microorganisms, potentially harmful chemicals, and such), be packaged in sanitary containers, and contain no added sweeteners or chemical additives (other than flavors, extracts, or essences). Flavors, extracts, and essences may be added as long as they comprise less than 1% by weight of the final product. |
| Municipal Water | Municipal water is simply tap water from a public water supply. Even though the tap water is perfectly safe to drink, bottlers typically treat the water through additional purification processes. If the water has not been significantly treated, then the FDA requires the label to state that the water is from a municipal source. |
| Purified Water | Purified water has gone through at least one of several processes to meet the U.S. Pharmacopeia (USP) definition of "purified." These processes can include distillation, deionization, and reverse osmosis. |

## What's Really in the Bottle? Data Table

Locate the requested information on the bottled water labels on the following pages and record the information here. Write a question mark if the information is not specifically given on the bottle label. Write inferences based on other parts of the label in brackets like this: [your inference].

| Label | Brand Name | Image on Label | Type of Water | Purification Treatment | Additives | Source of Water |
|---|---|---|---|---|---|---|
| Exhibit A | | | | | | |
| Exhibit B | | | | | | |
| Exhibit C | | | | | | |
| Exhibit D | | | | | | |
| Exhibit E | | | | | | |
| Exhibit F | | | | | | |
| optional bottled water | | | | | | |
| optional bottled water | | | | | | |
| optional bottled water | | | | | | |
| optional bottled water | | | | | | |

Take-Home Activity: What's Really in the Bottle?

# What's Really in the Bottle? Label Exhibits

**CRYSTAL MIST** is triple filtered for purity, using state of the art reverse osmosis treatment, and enhanced with minerals for a refreshing, pure taste. **CRYSTAL MIST** is water at its best.

**PURIFIED WATER**
Enhanced With Minerals for a Pure, Fresh Taste

*a product of The CRYSTAL MIST Company, Crystal River, FL*

20 FL OZ (1.25 PT) 591 mL

**Nutrition Facts**
Serv. Size 8 fl oz (240 mL)
Servings 2.5
Amount Per Serving
Calories 0
% Daily Value*
Total Fat 0g 0%
Sodium 0mg 0%
Total Carb 0g 0%
Sugars 0g
Protein 0g
*Percent Daily Values are based on a 2,000 calorie diet.

INGREDIENTS: PURIFIED WATER, MAGNESIUM SULFATE, POTASSIUM BICARBONATE, POTASSIUM CHLORIDE.

**Exhibit A**

---

Mineral Composition p.p.m. (mg/l)

| | | | |
|---|---|---|---|
| Calcium | 80 | Bicarbonates | 322 |
| Magnesium | 22 | Sulfates | 12 |
| Silica | 15 | Chlorides | 3 |
| | | Nitrates | 2 |

pH = 7.3
Dissolved solids = 312 p.p.m. (mg/l)

Bottled at source in Romania.

CA CASH REFUND

**bistra**
Natural Spring Water
*from the Carpathian Mountains*

**Bistra is a gift of nature.** Filtered through the Carpathian Mountains in the vicinity of world-famous mineral water health spas, Bistra is a pure and delicious part of each day.

**Nutrition Facts**
Serv. Size 8 fl oz (240 mL)
Servings 3
Amount Per Serving
Calories 0
% Daily Value*
Total Fat 0g 0%
Sodium 0mg 0%
Total Carb 0g 0%
Sugars 0g
Protein 0g
Calcium 2%
*Percent Daily Values are based on a 2,000 calorie diet.

25.3 FL OZ (0.79 QUART) 750 mL

**Exhibit B**

---

**BOTTLED IN FRANCE**

BOTTLED ONLY AT THE SOURCE. THE CARBONATION COMES FROM A NATURAL SOURCE BENEATH THE SPRING.

**BONNE EAU**

NATURALLY CARBONATED MINERAL WATER DISTRIBUTED BY BEST OF FRANCE, NEW YORK, N.Y.

**NATURAL SPARKLING MINERAL WATER**
LOW MINERAL CONTENT

**Exhibit C**

---

BOTTLED AT THE SOURCE

**Nutrition Facts**
Serv. Size 8 fl oz (240 mL)
Servings 2.5
Amount Per Serving
Calories 0
% Daily Value*
Total Fat 0g 0%
Sodium 0mg 0%
Total Carb 0g 0%
Sugars 0g
Protein 0g
*Percent Daily Values are based on a 2,000 calorie diet.

**SoPure**

SOPURE Spring Water gently flows from our own deep protected source beneath the rolling hills of Southern Michigan. Through nature's own filtration process, SOPURE Water slowly rises to the surface, both fresh and delicious.

TOTAL DISSOLVED SOLIDS: 140.0 P.P.M
ALWAYS BOTTLED AT THE SOURCE TO MAINTAIN THE QUALITY AND FRESHNESS OF OUR NATURAL SPRING WATER.

CA CASH REFUND

25.3 FL OZ (0.79 QUART) 750 mL

**Exhibit D**

# SEER
## NATURAL ARTESIAN WATER

**SEER Natural Artesian Water** begins as rainfall that filters into an aquifer deep beneath volcanic highlands and unspoiled tropical forests.

*Enjoy the pure, refreshing taste.*

Bottled in Tahiti, the largest island of French Polynesia, located in the southern Pacific Ocean.

Bottled at Source

Nutrition Facts
Serv. Size 8 fl oz (240 mL)
Servings 2.5

| Amount Per Serving | |
|---|---|
| Calories 0 | |
| | % Daily Value* |
| Total Fat 0g | 0% |
| Sodium 0mg | 0% |
| Total Carb 0g | 0% |
| Sugars 0g | |
| Protein 0g | |

*Percent Daily Values are based on a 2,000 calorie diet.

NON-CARBONATED

20 FL OZ
(1.25 PT) 591 mL

**Exhibit E**

# Yukon falls

## PREMIUM DRINKING WATER

Yukon Falls is a doubly purified water using carbon-activated ultra filtration and reverse osmosis. The water is then aerated to produce a refreshing taste. We remineralize with Calcium, Magnesium, and Potassium to give Yukon Falls a pure, natural taste.

Sodium Free

**1 GAL.**
(128 Fl. OZ.)

Nutrition Facts
Serv. Size 8 fl oz (240 mL)
Servings 3

| Amount Per Serving | |
|---|---|
| Calories 0 | |
| | % Daily Value* |
| Total Fat 0g | 0% |
| Sodium 0mg | 0% |
| Total Carb 0g | 0% |
| Sugars 0g | |
| Protein 0g | |
| Calcium 2% | |

*Percent Daily Values are based on a 2,000 calorie diet.

Source: Approved Municipal Source.
Yukon Falls, Middletown, Ohio

**Exhibit F**

*No fair peeking at this answer key until you've completed the data table.*

## Answer Key for What's Really In the Bottle? Data Table

| Label | Brand Name | Image on Label | Type of Water | Purification Treatment | Additives | Source of Water |
|---|---|---|---|---|---|---|
| Exhibit A | Crystal Mist | water drop | purified water | filtration, reverse osmosis | magnesium sulfate, potassium bicarbonate, potassium chloride | ? |
| Exhibit B | Bistra | mountains | spring water | ? [none] | ? [none] | Carpathian Mountains, Romania |
| Exhibit C | Bonne Eau | citrus fruit, glass of ice water | sparkling mineral water | ? [none] | ? [none] | spring, France |
| Exhibit D | So Pure | water pouring from bottle | spring water | ? [none] | ? [none] | spring, southern Michigan |
| Exhibit E | Seer | woman by waterfall, flower, and map of islands | artesian water | ? [none] | ? [none] | artesian well or spring, Tahiti |
| Exhibit F | Yukon Falls | waterfall | purified drinking water | carbon filtration, reverse osmosis | calcium, magnesium, potassium | municipal source, Middletown, Ohio |

**Take-Home Activity: What's Really in the Bottle?**

# The Amazing Water Maze

Take a journey through the amazing water maze!

A: Start your journey at entry ①.

B: Follow the pathway until you reach the exit marked ①.

C: Imagine you are collecting water at exit ①. What water type would you put on the bottle label? A hint is shown along the pathway. Write your answer in the Bottled Water Label chart, below right.

D: Repeat A–C, starting at entries ② ③ ④ ⑤ ⑥.

(One answer is done for you, and all of the answers are at the bottom of this page. No fair peeking until you're done!)

Stumped? Take a look at the "What's Really in the Bottle? Fact Sheet."

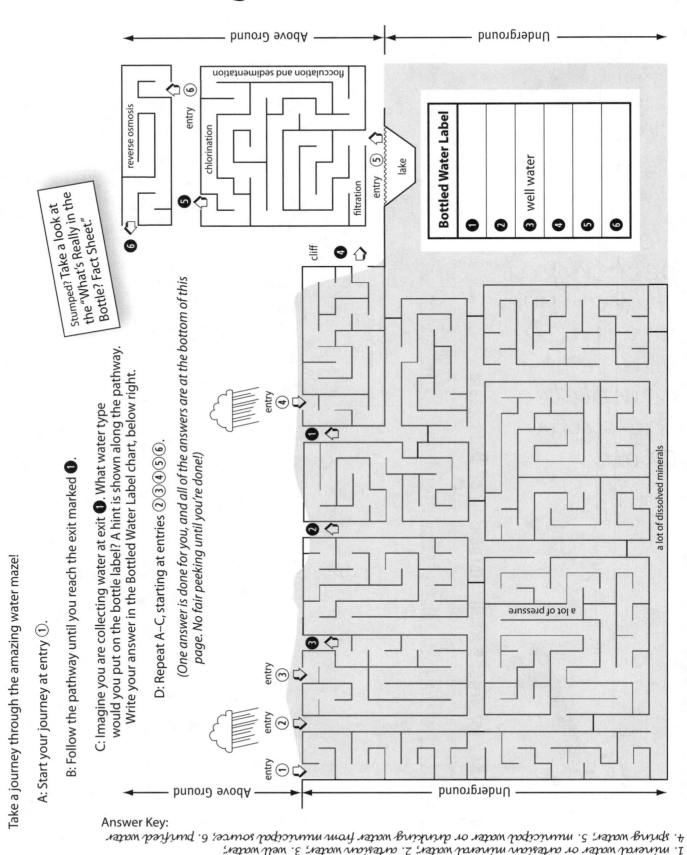

Answer Key:
1. mineral water or artesian mineral water; 2. artesian water; 3. well water; 4. spring water; 5. municipal water or drinking water from municipal source; 6. purified water

Page 84
© 2007 Terrific Science Press™
The publisher takes no responsibility for any use of any materials or methods described in this monograph, nor for the products thereof. This publication was made possible by Grant Number 1 R25 RR16301-01A1 from the National Center for Research Resources (NCRR), a component of the National Institutes of Health (NIH). Its contents are solely the responsibility of the authors and do not necessarily represent the official views of NCRR or NIH.

Take-Home Activity: The Amazing Water Maze

# TOPIC 3

## HEALTHY AIR

### List of Activities

- ✓ Look at One Part per Million .................................................. 86
  *Create a one part per million (1 ppm) solution to learn that 1 ppm is not necessarily an insignificant quantity.*

- ✓ Balloon Challenge ................................................................... 88
  *Explore the fact that air takes up space.*

- ✓ Search for One Part per Million ............................................. 90
  *Discover how easy it is to detect one part per million (1 ppm).*

- ✓ Humidity Detector .................................................................. 91
  *Discover how humidity varies from place to place.*

- ✓ How Much Oxygen Is in Air? .................................................. 93
  *Determine the amount of oxygen in air by observing the oxidation of iron.*

- ✓ Pour a Gas ................................................................................ 96
  *Observe the presence of an invisible gas.*

- ✓ Pour More Gas ......................................................................... 98
  *Learn how to detect an invisible gas.*

- ✓ Stirring Indoor Air ................................................................. 100
  *Observe the effects of both tightly sealed and well-ventilated buildings on indoor air pollution.*

- ✓ Trapping Particulates ........................................................... 102
  *Explore the ability of vacuum bags to trap fine particles such as biological contaminants.*

- ✓ Cartesian Diver ...................................................................... 104
  *Investigate the compressibility of air.*

- ✓ Take-Home Activity: Growing Mold .................................... 106
  *Learn what conditions help mold to grow.*

# Look at One Part per Million

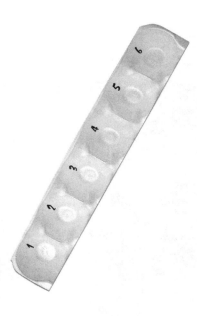

## Overview

Environmental contaminants, such as pollutants and greenhouse gases, are often measured in parts per million (ppm). One ppm sounds like a pretty tiny amount, but let's see how easy it is to make a 1 ppm concentration.

## Materials

- ✓ egg carton (prepared by leader) or 6 small white or clear containers (such as plastic cups)
- ✓ self-stick labels or waterproof marker
- ✓ piece of white paper (if clear containers are used)
- ✓ 6 droppers or disposable pipets
- ✓ water
- ✓ blue or green liquid food color

## Procedure

1. Number six depressions in the egg carton from 1 to 6. Alternately, number six containers and place them side by side (on white paper if they are clear).

2. Use a dropper to put 9 drops of water into each container. Be sure to hold the dropper vertically (straight up and down) to keep the size of the drops consistent.

3. Use a clean, dry dropper to put 1 drop of food color into container 1. Stir with the dropper to mix well. The concentration of this solution is 1 drop of food color per 10 drops of total solution (1 part per 10). This is recorded for you in the data table at the end of this activity. Record whether the color is still visible at this concentration.

4. Use a clean, dry dropper to transfer 1 drop from solution container 1 to container 2. Stir to mix well. This is 1 part food color per 100 parts of solution. Record this dilution in the data table and whether the color is still visible.

## Camper Notebook

5. Continue this process for the remaining containers. The last dilution is 1 part food color per 1,000,000 parts of solution, or 1 ppm.

**Q1:** At what dilution does the solution first appear colorless?

|  | Container | | | | | |
|---|---|---|---|---|---|---|
|  | 1 | 2 | 3 | 4 | 5 | 6 |
| concentration | 1 part per 10 | | | | | |
| Is color visible? | | | | | | |

Healthy Air—Look at One Part per Million

# Balloon Challenge

## Overview

Is an empty bottle really empty? We usually describe a bottle that does not contain a liquid or solid as being empty, as if the air inside does not exist. But air is present and has properties of its own. In this activity, you will examine one of these properties.

## Materials

✓ balloon
✓ empty and clean 1-L or 2-L soft-drink bottle
✓ soft-drink bottle prepared by leader

## Procedure

1. Push a deflated balloon into the first soft-drink bottle and stretch the open end of the balloon back over the mouth of the bottle as shown. Now blow into your balloon.

**Q1:** What happens?


2. Take your balloon out of the bottle and blow into the balloon.

**Q2:** What happens?


**Q3:** Can you explain the difference between what happens in step 1 and what happens in step 2?


3. Push your deflated balloon into another bottle (one that your leader gives you) just like you did in step 1. Now blow into your balloon.

## Camper Notebook

**Q4:** What happens? Can you explain why this happens? What do the results of steps 1–3 tell you about what's in the bottle?

**4.** With the setup used in step 3, hold your finger over the hole on the side of the bottle and then blow into your balloon.

**Q5:** What happens and why?

**5.** With the setup used in step 3, leave the bottle's hole open and blow into your balloon. With your mouth still over your balloon's opening, place a finger tightly over the bottle's hole. Now remove your mouth but keep your finger over the hole.

**Q6:** What happens and why?

Healthy Air–Balloon Challenge

# Search for One Part per Million

## Overview

Scientists measure air pollutants using special instruments that are sensitive enough to detect part-per-million (ppm) concentrations. How easy is it to detect just 1 ppm without these special intruments?

## Materials

- ✓ jar containing one dark-colored particle in white salt (prepared by leader)
- ✓ watch or clock with second hand
- ✓ bottle of water containing one piece of glitter (prepared by leader)

## Procedure

1. Time and record at left how long it takes you to find the dark-colored particle that is "one in a million." If you'd like, shake the jar and time your search several more times.

2. Now look at the bottle of water. There is one piece of glitter in the water at a concentration of 1 ppm.

**Q1:** Does the piece of glitter disappear in the water or is it easily visible? Why?

**Q2:** How can the glitter be removed?

| I found the particle in how much time? |
|---|
|  |
|  |
|  |
|  |

# Humidity Detector

## Overview

Dry air is mainly nitrogen, oxygen, and argon. Water vapor is another gas that is present in the air. Air that contains too much moisture may promote mold and mildew growth, which can cause health problems. Air that is too dry may cause uncomfortably dry eyes and skin. Is it possible to detect the relative amount of moisture in the air by a color change? Find out by making a humidity detector.

## Materials

- ✓ round white filter paper about 8 inches (20 cm) in diameter
  - ☞ *If a cone-shaped or basket-type coffee filter is used, not as many folds are needed to create the wedge shape in step 1.*
- ✓ green pipe cleaner or chenille stem about 12 inches (30 cm) in length
- ✓ 2 green tissue paper rectangles about 4 inches × 6 inches (10 cm × 15 cm)
- ✓ scissors
- ✓ 10% aqueous cobalt chloride solution prepared by leader
  - ⚠ *Cobalt chloride is harmful if taken internally. Do not taste the cobalt chloride solution. Wash skin well if contact occurs. Wear goggles when handling the solution.*
- ✓ shallow container
- ✓ place to hang humidity detector to dry (such as clothes drying rack or clothesline)
- ✓ clothespin
- ✓ hair dryer
- ✓ fine-mist sprayer bottle filled with water
- ✓ goggles

## Procedure

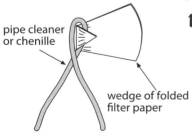

pipe cleaner or chenille
wedge of folded filter paper

1. Fold the round filter paper in half about three times to make a wedge shape. Fold the wedge about 1 inch (3 cm) from the pointed end. Bend the center of the pipe cleaner over the fold as shown and twist the pipe cleaner tightly three times.

## Camper Notebook

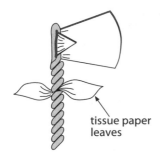
tissue paper leaves

2. Place the center of the tissue paper rectangles into the pipe cleaner and continue twisting the stem to the end as shown.

3. Put on goggles. Hold the "flower" upside down and dip the filter paper briefly into the 10% cobalt chloride solution prepared by your leader.

4. Hang the flower upside down by the stem and dry it with a hair dryer.

**Q1:** What color is the cobalt chloride when the cobalt chloride is in solution and first applied to the filter paper?

**Q2:** What color is the cobalt chloride when the paper is dry?

5. Open the blossom of the flower, spray the flower with water using a mist sprayer, and observe what happens. Dry the flower again like you did in step 4.

6. Place the flower in different locations, such as near a heating vent, ceiling, and open window. Try putting the flower in places such as a closet, basement, attic, and crawl space. In the data table, record each location and the flower's color at that location. Based on what you observed about the flower's color in steps 4 and 5, decide and record the relative amount of moisture in each place (for example, high moisture, moderate moisture, dry).

⚠ *Be sure to keep the flower away from small children and pets.*

| Location | Flower Color | Relative Amount of Moisture |
|---|---|---|
|  |  |  |
|  |  |  |
|  |  |  |
|  |  |  |
|  |  |  |
|  |  |  |
|  |  |  |
|  |  |  |
|  |  |  |

# How Much Oxygen Is in Air?

## Overview

Air is made up of many gases. Oxygen, one of the gases in air, is essential for living things. How much of the air is oxygen? In this activity, you will perform a chemical reaction and take measurements to find out.

## Materials

- ✓ about 20 mL 0.5 M acetic acid (prepared by leader)
- ✓ 50-mL beaker
- ✓ 1 g fine or medium-fine steel wool
- ✓ balance capable of measuring 0.1 g
- ✓ paper towels
- ✓ test tube about 15 cm (6 inches) long
- ✓ metric ruler at least 15 cm (6 inches) long
- ✓ tape that will stick when submerged in water
- ✓ 400-mL beaker or a cut-off 1-L clear soft-drink bottle
- ✓ water
- ✓ stirring rod
- ✓ watch or clock with second hand

## Procedure

1. Pour about 20 mL of the 0.5 M acetic acid into a 50-mL beaker. (This will be used to activate the steel wool's surface.)

2. Immerse about 1 g steel wool in the acetic acid solution for about 1 minute. Remove the steel wool and press the excess liquid out with paper towels. (Be careful because the steel wool is sharp.) Rinse your hands with water after handling the acetic acid solution.

3. Tape a metric ruler to the side of a test tube so that the zero mark is at the mouth (open end) of the test tube as shown. Measure the length of the test tube. To compensate for the tube's rounded bottom, measure to only halfway up the bottom's curve as shown. Record the length of the test tube in the data table on the next page. Be sure to include the units in your measurement.

Healthy Air—How Much Oxygen Is in Air?

## Camper Notebook

4. Fill a 400-mL beaker about ²/₃ full of water.

5. Carefully pull the steel wool apart to increase its surface area. (Be careful because steel wool is sharp.) Measure the mass of the steel wool and record the information in the data table. Use a stirring rod to insert the steel wool so that it stays at least halfway up the test tube. Wash your hands.

6. Quickly turn the test tube over and place it into the beaker so that the test tube's mouth stays below water level. Record the starting time when the test tube is inserted into the water. Measure and record the water level in the test tube at the starting time. When taking the water level readings, keep the test tube inverted in the water but line up the water level inside the test tube with the water level outside the test tube.

7. Every 3 minutes, record the time and the water level in the test tube. After 15 minutes, the readings can be taken every 5 minutes. Take the last reading at 40 minutes (if time permits).

| Length of test tube: | |
|---|---|
| Mass of steel wool before the reaction: | |
| **Time** | **Water Level in Test Tube** |
| (starting time) | |
| (after 3 minutes) | |
| (after 6 minutes) | |
| (after 9 minutes) | |
| (after 12 minutes) | |
| (after 15 minutes) | |
| (after 20 minutes) | |
| (after 25 minutes) | |
| (after 30 minutes) | |
| (after 35 minutes) | |
| (after 40 minutes) | |

**Q1:** What everyday process is happening in the test tube?

# Camper Notebook

**Q2:** Is all of the steel wool reacted in the test tube? If not, what other reactant may be used up in the test tube?

**Q3:** Do you think that the steel wool is gaining mass or losing mass? How could you check?

**Q4:** What else do you need to consider when evaluating the steel wool's mass?

**8.** Plot a graph showing water level in the test tube versus time. Be sure to label each axis and include units.

**Water Level Versus Time**

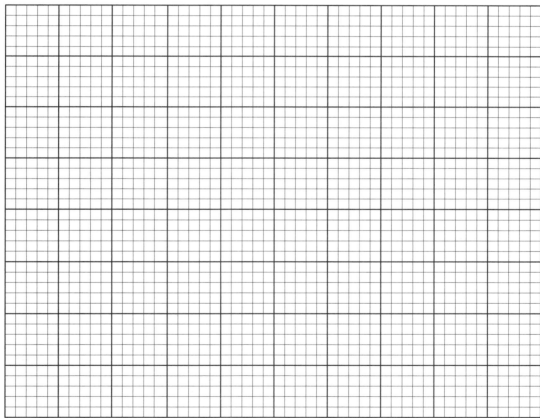

**9.** Do this calculation to determine the percentage of oxygen in air:

$$\frac{\text{final water level in test tube}}{\text{test tube length}} \times 100 = \% \text{ oxygen}$$

Healthy Air—How Much Oxygen Is in Air?

# Pour a Gas

## Overview

The concentration of a certain colorless, odorless gas in our atmosphere has increased from about 280 parts per million (ppm) in the year 1750 to about 360 ppm in 2000. What is this gas? The following activity shows ways to detect its presence.

## Materials

- ✓ birthday candle
- ✓ scissors
- ✓ foil cupcake liner (discard paper separators)
- ✓ aluminum foil or clay
- ✓ graduated beaker or liquid measuring cup
- ✓ at least ⅓ cup (75 mL) vinegar
- ✓ 2-L plastic soft-drink bottle
- ✓ matches (for adult use only)

⚠ *There is a danger of fire even with small birthday candles. The work area must be clear of combustible materials. Loose clothing is discouraged and long hair must be tied back.*

- ✓ funnel (a cone made of paper will also work)
- ✓ at least 1 teaspoon (5 mL) baking soda
- ✓ teaspoon or other volume measure

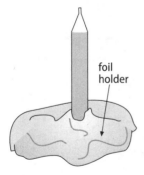

foil holder

## Procedure

1. Cut a birthday candle a little bit shorter than the height of a foil cupcake liner. Use a bit of foil or clay to make a holder so the candle stands in the center of the foil cup. The top of the candle's wick should be below the rim of the foil cup.

2. Pour about ⅓ cup (75 mL) vinegar into a soft-drink bottle. Although the bottle now contains a little vinegar, it is mostly filled with the gases that make up air.

**Q1:** Do you think the gas in the bottle will extinguish (put out) a candle flame?

## Camper Notebook

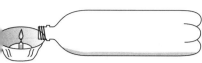

**3.** Light the candle. Slowly tip the bottle as if pouring the gas into the foil cup. Stop tipping before any of the liquid comes out of the bottle.

**Q2:** Does the candle go out? If so, why?

**4.** Tip more to pour a few drops of vinegar into the foil cup around the candle. (Do not let the vinegar drops hit the candle flame itself.)

**Q3:** Does the candle go out? If so, why?

**5.** Using a funnel or paper cone, add about 1 teaspoon (5 mL) baking soda into the bottle containing the vinegar. Swirl the bottle to make sure the liquid and powder mix well.

**Q4:** What happens when baking soda is added to vinegar?

**6.** When the reaction dies down, try tipping the bottle over the foil cup around the lit candle just as you did in step 3. (It may be necessary to keep pouring until a few drops of liquid fall into the foil cup. It may also help to squeeze the bottle to force out some of the gas.)

**Q5:** Does this gas extinguish the flame? Based on your observations, what did you learn about the gas formed by combining vinegar and baking soda?

**Q6:** What do you think would happen if your house was completely filled with this colorless, odorless gas instead of air?

# Pour More Gas

## Overview

In "Pour a Gas," you made a colorless, odorless gas and used it to put out a candle. In this activity, you will try to visually detect the gas using litmus paper.

## Materials

- ✓ small jar with a lid
- ✓ water
- ✓ blue litmus paper
- ✓ club soda
- ✓ small cup
- ✓ about ⅓ cup (75 mL) vinegar
- ✓ graduated beaker or liquid measuring cup
- ✓ 1-L or 2-L plastic soft-drink bottle
- ✓ about 1 teaspoon (5 mL) baking soda
- ✓ teaspoon or other volume measure
- ✓ funnel (a cone made of paper will also work)

## Procedure

1. Fill a small jar about one-quarter full of water.
2. Dip one end of a piece of blue litmus paper into the water.

**Q1:** What color is the litmus paper when it is wetted by water? What does this indicate?

**Remember...**
- The sample is acidic if blue litmus paper turns red.
- The sample is not acidic if blue litmus paper stays blue.

3. Pour a little club soda into a small cup. Dip one end of a new piece of blue litmus paper into the club soda.

**Q2:** What color is the litmus paper when it is wetted by club soda? What does this mean?

# Camper Notebook

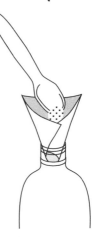

4. Pour about ⅓ cup (75 mL) vinegar into a soft-drink bottle. Use a funnel to add about 1 teaspoon (5 mL) baking soda into the bottle as shown. Swirl the bottle to make sure the liquid and powder mix well.

5. When the reaction in the bottle slows down, pour some of the gas from the bottle into the jar of water just as you poured the gas into the foil cup containing the lit candle in step 3 of "Pour a Gas." Stop pouring before any of the liquid comes out of the bottle. When you have poured the gas into the jar, seal the jar tightly and shake it vigorously.

6. Now test to see if some of the colorless, odorless gas you made in the soft-drink bottle has dissolved in the water. Open the jar and dip one end of a new piece of blue litmus paper into the water.

**Q3:** What color is the litmus paper when dipped into this solution? What does this mean?

**Q4:** How does the color of the litmus paper from the jar (in step 6) compare to the color of the litmus paper from the cup containing club soda (in step 3)?

**Q5:** What gas is used to make the "fizz" in club soda? (If you don't know, make a guess.)

**Q6:** The gas you made in step 4 by reacting vinegar and baking soda is colorless and odorless, but do you think it is tasteless? Why or why not?
⚠️ *Do not taste any of the materials in this experiment.*

# Stirring Indoor Air

## Overview

Indoor air pollution is a problem a lot of us never think about, but one that is an increasing concern to many government health agencies. U.S. Environmental Protection Agency (EPA) studies show that the indoor levels of many pollutants can be 2–5 times, and sometimes more than 100 times, higher than outdoor levels. In this activity, you will use flour to represent the spread of air pollutants inside both tightly sealed and well-ventilated model cardboard houses.

## Materials

- ✓ 2 small identical cardboard boxes with lids
- ✓ scissors
- ✓ ruler
- ✓ clear plastic food wrap
- ✓ masking tape
- ✓ paper to make a funnel
- ✓ 6 spoonfuls of flour
- ✓ plastic spoon
- ✓ clear plastic storage tub
- ✓ turkey baster

## Procedure

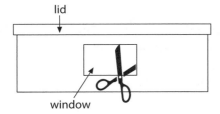

1. Make two "houses" from cardboard boxes by cutting identically sized windows in the two opposite sides of each box as shown. Cut a hole about the size of a dime in the center of each lid.

2. On one box, tightly tape plastic food wrap over both windows (so that they are well-sealed all the way around). Tightly seal all seams of the box with tape (including around the edges of the lid). This box represents a tightly sealed house. The other box represents a well-ventilated house.

# Camper Notebook

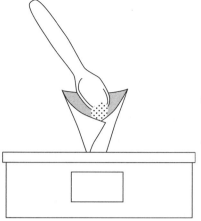

3. Place the tip of a paper funnel through the hole in the lid of one of the boxes. Put three spoonfuls of flour in the funnel so flour drops inside the box. This flour represents pollution that is already present inside the house.

4. Repeat step 3 for the second box. Put both boxes side by side in a plastic tub.

5. Put the tip of the turkey baster into the hole in the lid of one of the boxes. Squeeze the bulb of the turkey baster firmly several times while pointing it around the inside of the box to stir up the flour. The air coming from the turkey baster represents the air circulating inside the house. Observe what happens to the flour. You can remove the box lids to observe inside.

6. Repeat step 5 with the other box.

**Q1:** Where did the flour inside each of the "houses" go when stirred around by air?

**Q2:** Which of these houses became more polluted after the air circulated?

**Q3:** What could you do to reduce the pollution in the more polluted house?

**Healthy Air—Stirring Indoor Air**

# Trapping Particulates

## Overview

Particulate matter (such as dust and soot) and biological contaminants (such as mold, pollen, animal dander, and bacteria) get into our homes and can cause poor indoor air quality. Can we reduce or eliminate these contaminants by ordinary cleaning methods like vacuuming?

## Materials

- ✓ two or more different types of disposable vacuum cleaner bags
- ✓ scissors
- ✓ packages from the vacuum cleaner bags used in this activity
- ✓ sales slip listing the cost of each package of vacuum cleaner bags
- ✓ fine powder (such as cornstarch, talcum powder, cinnamon, or activated charcoal)
- ✓ teaspoon or other volume measure
- ✓ tape
- ✓ paper having a color that contrasts with the fine powder
- ☞ *For example, use black paper if testing with cornstarch or talcum powder and white paper if testing with activated charcoal or cinnamon.*

## Procedure

1. Cut the attachment opening off each vacuum cleaner bag so that you are left with bag-like filters having one opened end. (If the vacuum bag is long and the attachment opening is near the middle, both the top and bottom of the bag can be used. Share the other half with another person or group.)

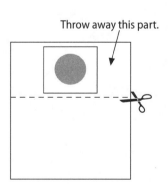

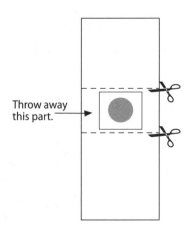

# Camper Notebook

**2.** In the data table at the end of this activity, record the type of powder and the brands of bags you will test. List any performance claims printed on the package (such as "stress tested," "allergen reduction," or "HEPA filter") and the cost per bag of each brand. Divide the cost of the package by the number of bags in the package to get the cost per bag.)

**3.** Look closely at the construction of each brand of bag, including its thickness and the material it's made of. Describe each bag's construction in the data table.

**4.** Add 1 teaspoon (5 mL) fine powder to each bag. Fold the tops of the bags down twice to make the seals powder tight. Tape the folds closed.

**5.** Count the taps as you tap each bag repeatedly on the table over a piece of paper. As soon as the powder works its way through the bag and onto the paper, stop tapping and record the number of taps in the data table. If no powder is observed after 150 taps, record "trapped after 150."

**Q1:** Which bag did the best job of trapping very fine powder?

**Q2:** Considering the cost and filtering abilities of the different bags, which bag would you buy and why?

| Powder Tested: | | | | |
|---|---|---|---|---|
| **Brand of Vacuum Bag** | **Performance Claim** | **Cost per Bag** | **Bag Construction** | **Number of Taps** |
| | | | | |
| | | | | |
| | | | | |
| | | | | |
| | | | | |
| | | | | |

Healthy Air–Trapping Particulates

# Cartesian Diver

## Overview

Just like solids and liquids, gases have their own properties. In this activity, you will investigate a property of gas while observing the movement of a Cartesian diver in water.

## Materials

- ✓ disposable polyethylene graduated pipet
- ✓ scissors
- ✓ hex nut
- ✓ cup at least 4 inches (10 cm) tall
- ✓ water
- ✓ 1-L soft-drink bottle with cap
- ✓ piece of glitter

## Procedure

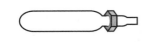

1. Make a diver by cutting off the graduated stem of a disposable pipet as shown. Be sure to leave about ½ inch (1½ cm) of the stem on the pipet below the bulb.

2. Screw a hex nut on the remaining stem of the pipet. Turn hard to slightly thread the hex nut onto the plastic until the hex nut is up tight against the bulb of the pipet as shown. (If too much of the stem of the pipet protrudes, snip it off with scissors.)

3. Fill the cup with water. Hold the pipet so the weighted end is down. Squeeze the sides of the weighed pipet and allow some water into the open end of the pipet. Release the pipet and observe whether it sinks or floats in the cup of water. Adjust the amount of water in the bulb so that it floats in the cup of water as shown.

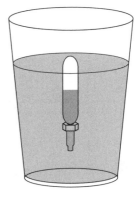

# Camper Notebook

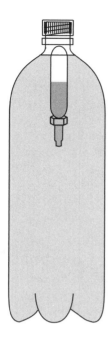

4. Fill the soft-drink bottle nearly to the top with water. Add one piece of glitter. This piece of glitter represents a pollutant present in water at 1 part per million (ppm).

5. Carefully remove the pipet (diver) from the cup of water without losing any water from inside the diver. Place the diver in the bottle. If the diver does not float near the top of the bottle as shown, remove the diver and adjust the amount of water in the bulb.

6. Screw the cap on the bottle tightly. Press or squeeze the sides of the bottle and observe what the diver does. Then, observe the diver when the bottle is released.

**Q1:** What happens to the diver in step 6?

**Q2:** Describe what happens to the volume of gas (air) inside the diver as you squeeze the bottle (increase the pressure on the bottle).

**Q3:** Do you think it is possible for the piece of glitter to end up inside the diver? Why or why not?

# Growing Mold
## What conditions help mold to grow?

*Individual mold spores are invisible to the naked eye. Given the proper conditions, however, they will germinate and grow into large colonies of mold.*

FYI: This activity requires about a week to complete.

Perhaps you've been disgusted by mold growing on bread, cheese, or other food. Some molds are dangerous, but most are relatively harmless. Let's try to grow mold under safe conditions.

## What You'll Need:
- 2 or more sandwich-sized zipper-type plastic bags
- self-stick labels or permanent marker
- 2 or more paper towels
- water
- 2 or more saltine crackers

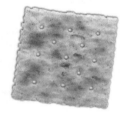

## What You'll Do:

>> 1. Label one zipper-type plastic bag "dry" and another bag "moist."

>> 2. Fold a dry paper towel so it is the size of a saltine cracker. Place one cracker on the table and then place the dry, folded paper towel on top of the cracker. Slide both the paper towel and the cracker into the plastic bag labeled "dry" so that the label blocks the view of the paper towel and not the view of the cracker. Zip the bag.

>> 3. Dampen another paper towel, but do not make it soaking wet. Repeat step 2 with the damp paper towel, another cracker, and the plastic bag labeled "moist." Zip the bag.

>> 4. Place both bags out of the way on a table or counter at room temperature.

>> 5. If you'd like, you can prepare and store one or more additional crackers any way you want to test other conditions that might grow mold.

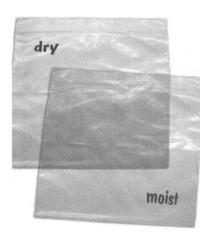

>> **6.** In the data table, record the start date and your observations about mold growth. For the next week, do not open the bags but look through the bags and check each cracker for mold growth at about the same time each day. Record your observations in the data table.

 *Do not open the bags at any time during the week or at the end of this activity because mold spores should not be released into the air in high numbers. When you are done with the activity, throw away all bags without opening them.*

| Growing Mold Data Table | | | | |
|---|---|---|---|---|
| | Observations | | | |
| Date | Dry Cracker in Bag | Moist Cracker in Bag | Other: | Other: |
| | | | | |
| | | | | |
| | | | | |
| | | | | |
| | | | | |
| | | | | |
| | | | | |
| | | | | |

## Questions to Consider:

- What conditions and what length of time are needed for mold to grow on a cracker?
- Where do you think the mold on the cracker comes from?
- How many types of mold do you think are present on the cracker? What is your evidence?
- Where in your home can you find conditions right for mold growth?

## Variations You Can Do:

Try growing molds on different types of cracker, cookie, bread (with and without preservatives), or grain products. Observe the difference in types and quantities of mold growth on these products.

## What's the Deal?

In this activity, you may discover that mold needs a moist environment, moderate temperatures (not too hot and not too cold), and a nutrient source to grow. (The cracker is the nutrient source.) After a week, your moist cracker should have interesting mold growth. Although different kinds of molds can only be positively identified under a microscope, cottony masses are usually species of *Rhizopus* (black bread mold). Bluish-green to green molds are usually *Penicillium* or *Aspergillus*. Black to brown-black molds can be species of *Aspergillus, Alternaria, Cladosporium,* or *Stachybotrys*. Reddish or pink molds can be species of *Fusarium*.

Mold spores are floating around in air all of the time. They also settle on surfaces like the table where you laid the cracker right before putting it in the bag. If conditions are good for growth, the spores will produce mold. Damp basements, bathrooms, and ventilation systems are often ideal locations for mold to grow in our homes.

Cracker in plastic bag on first day (left) and after 7 days (right)

# Topic 4

## Healthy Skin

### List of Activities

- ✓ Hydrophobic Art ........................................... 110
  *Explore hydrophobic paint to learn how skin serves as a protective layer for the outside of the body.*

- ✓ Visible Light Challenge .................................. 112
  *Investigate the relative energy of visible light using different colors of light.*

- ✓ How Sensitive Is Your Skin? ........................... 113
  *Measure the sensitivity of different parts of the hand.*

- ✓ Make and Test Lip Balm ................................. 115
  *Create and test several lip balm products to assess their sunscreen and waterproof properties.*

- ✓ Cover Up, Screen, or Block? ........................... 119
  *Discover the UV protection of certain clothing, sun products, and sunglasses.*

- ✓ Sunning Straws .............................................. 122
  *Use benzophenone and a UV light reaction to quantitatively measure the amount of UV protection that different sun protection products provide.*

- ✓ Suntan in a Bottle .......................................... 124
  *Investigate how a sunless tanning product reacts with different natural and synthetic fabrics.*

- ✓ A Look at Bleaching ....................................... 127
  *Observe the effects of bleach on fabric to mimic what skin bleaching products might do to skin.*

- ✓ Take-Home Activity: UV Detective Challenge ..... 130
  *Make a UV detection bracelet, then place the bracelet in different outdoor locations to test levels of UV radiation.*

# Hydrophobic Art

## Overview

Skin serves as a protective layer for the outside of the body. Skin allows some liquids out of the body (such as sweat), but it usually holds other body fluids inside. Skin is mostly hydrophobic, meaning that it is "water hating" (it lacks affinity for water). Notice that rain drops run off the outside of our skin. Explore another hydrophobic material in the following activity as you decorate a pencil to take home.

## Materials

- ✓ newspaper and paper towels
- ✓ clear plastic 1- or 2-L soft-drink bottle (prepared by leader)
- ✓ water
- ✓ toothpicks
- ✓ 2 different colors of Testors brand oil-based enamel model paint

   ⚠ *Work in a well-ventilated area. Enamel paint is flammable, so avoid sparks and open flames. Avoid getting paint on the skin. If you do, use an oil-paint solvent to remove the paint and then wash the skin with plenty of soap and water.*

- ✓ masking tape
- ✓ wooden pencil or wood cut-out
- ✓ spring-type clothespin
- ✓ oil-paint solvent (such as turpentine) for emergency clean-ups

   ⚠ *Handle the solvent as directed on its package label.*

## Procedure

1. Lay newspaper over the work area. Fill the cut-off soft-drink bottle with water as shown.

2. Using the precautions listed above for handling enamel paint, take a clean toothpick and drop about 4–5 drops of one color of enamel paint on the water. Notice what happens to the paint.

**Q1:** Does the paint sink or float? Does it mix with the water?

## Camper Notebook

3. Add 4–5 drops of a second color of enamel paint using a clean toothpick. Then, use this toothpick to slowly swirl the paint into a marbled pattern.

4. Dip one end of another clean toothpick about halfway into the marbling mixture. Look through the side of the bottle when you do this. Then remove the toothpick and notice the effect on the toothpick.

**Q2:** What do you observe as the toothpick is dipped into the water?

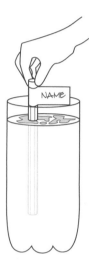

5. Write your name on a piece of tape and attach the tape to the eraser end of a pencil as shown.

6. Holding the pencil by the eraser, dip the pencil into the marbling mixture while slowly twirling the pencil as shown. (Twirling the pencil while dipping prevents paint from clumping together in one spot and produces a more even marbling effect.)

7. Remove the pencil from the water after it is dipped down to the label as shown. Clip a clothespin onto the taped end of the pencil. Prop the pencil over newspaper and paper towels by using the pencil's lead end and the clothespin's two open feet as shown. Allow the pencil to completely dry.

8. If another person needs to dip his or her pencil into the same container, repeat steps 2 and 3 to add more paint to the water as needed.

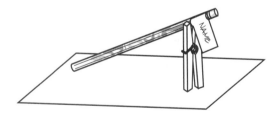

# Visible Light Challenge

## Overview

UV radiation is high in energy and therefore damaging to skin. Although we cannot see UV radiation, this activity uses visible light of different colors to introduce the idea that different wavelengths of light have different levels of energy. Which of the colors has the highest energy?

## Materials

- ✓ darkened room
- ✓ red, green, and blue light emitting diode (LED) mini flashlights
- ✓ phosphorescent (glow-in-the-dark) vinyl sheet
- ✓ timer such as stopwatch or watch with second hand

## Procedure

1. In a darkened room, turn on either the green or blue LED mini flashlight. Hold the flashlight against the green side of the phosphorescent sheet so that the light is touching the sheet. Start timing and move the flashlight around on the phosphorescent sheet for 10 seconds. With the timer still ticking, turn off the flashlight and time how long the trail is visible on the phosphorescent sheet.

2. Record your results in the data table below.

3. Repeat steps 1 and 2 with the same phosphorescent sheet and the other two LED mini flashlights (one at a time).

**Q1:** Based on your results, which of the emitted lights do you think has the highest energy and which has the lowest energy? Why?

| LED Color | Phosphorescent Trail? (Yes or No) | How Long the Trail Lasted |
|---|---|---|
| | | |
| | | |
| | | |

# How Sensitive Is Your Skin?

## Overview

Some people who don't use sunscreens or sunblocks may think that they will "feel" when they have had enough sun. The problem with this theory is that sunburn has already happened by the time they feel it. Skin contains neurons that respond to the sense of touch. In this investigation, you will measure the sensitivity of different parts of the hand with a set of testing devices called von Frey hairs.

## Materials

- ✓ 5 different thicknesses of fishing line having a wide range of diameters
- ☞ *Transparent sewing thread may be used for a thin diameter.*
- ✓ scissors
- ✓ tape
- ✓ craft sticks
- ✓ (optional) blindfold

## Procedure

1. Make your own set of von Frey hairs with different thicknesses of fishing line. For each thickness, cut a length of fishing line about 5 cm (2 inches) long. Tape the end of the fishing line onto a craft stick at a 90° angle as shown. Label the craft stick with the diameter of the fishing line.

2. Working in pairs or small groups, decide on up to five parts of the hand that you want to test. Circle and number these locations in the hand diagrams at the end of this activity. Record your fishing line thicknesses in the data table at the end of this activity by listing the thinnest size to the thickest size.

3. Decide on the person whose hand will be tested. Blindfold that person or have him or her look away. Touch the thinnest fishing line to the first hand location until the fishing line bends. Do not move the fishing line along the skin. Ask the person if he or she feels anything and record the answer in the data table. Repeat the procedure with the same fishing line for all hand locations.

**Camper Notebook**

4. Repeat step 3 for all fishing line thicknesses. If time permits and if you want, test the hands of other group members and record their answers in the data table using a method that distinguishes one set of answers from another (such as different ink color, cursive writing, all capital letters, or something else).

5. Tactile detection threshold is the smallest amount of touch necessary for a person to feel. For example, imagine touching four different thicknesses of fishing lines (0.30 mm, 0.41 mm, 0.61 mm, and 0.89 mm) on the tip of an index finger. If the person only felt the 0.61 mm and 0.89 mm thicknesses, the tactile detection threshold would be 0.61 mm (the smallest diameter of line felt). In your data table, circle the tactile detection threshold for each hand location you tested (and for each person's hand you tested).

**Q1:** Look at the data table. If the detection threshold is different for different parts of the hand, describe the differences.

**Q2:** Where is the most sensitive part of the hand that you tested?

**FYI...**
Circle and number hand locations on these hand diagrams.

palm                back

| Write fishing line thicknesses across this row (thinnest to thickest). → | Was the touch felt with the following fishing line thickness? | | | | |
|---|---|---|---|---|---|
| location 1 | | | | | |
| location 2 | | | | | |
| location 3 | | | | | |
| location 4 | | | | | |
| location 5 | | | | | |

# Make and Test Lip Balm

## Overview

In this activity, you will make lip balm using oil of cinnamon as a flavor. You will also create test products having varying concentrations of oil of cinnamon and then test these products for sunscreen and waterproof abilities.

## Materials

- ✓ self-stick labels
- ✓ permanent marker
- ✓ 5 or 6 mini foil muffin cups
- ✓ lanolin
- ✓ balance capable of measuring 0.1 g or set of measuring spoons
  - ☞ *Measuring ingredients is more accurately done by mass with a balance rather than by volume with measuring spoons.*
- ✓ coconut oil
- ✓ beeswax
- ✓ electric food-warming tray
  - ⚠ *Use the warming tray carefully and with adult supervision.*
- ✓ toothpicks
- ✓ oil of cinnamon
  - ⚠ *Large amounts of oil of cinnamon can be irritating to skin.*
- ✓ dropper or disposable pipet
- ✓ 12 paper twist ties
- ✓ 12 UV detection beads of the same color
- ✓ waxed paper
- ✓ cup of water
- ✓ (optional) knife to break up the beeswax
- ✓ (optional) paper such as paper muffin cup
- ✓ (optional) small plastic knife, spatula, or bowl scraper
- ✓ (optional) towel to wrap around electric food-warming tray
- ✓ (optional) UV lamp
  - ⚠ *Looking directly at the UV light can cause eye damage.*

**Camper Notebook**

## Procedure

1. Label a mini foil muffin cup with your name and "LB" to stand for "lip balm." Add 1.0 g (¼ teaspoon) lanolin, 5.0 g (1¼ teaspoon) coconut oil, and 4.0 g (1 teaspoon) beeswax according to your leader's instructions.

2. Heat the muffin cup on an electric food-warming tray until all ingredients are melted. Stir occasionally with a toothpick.

3. Remove the muffin cup from the warming tray and add 2 drops oil of cinnamon. Stir. Allow the mixture to cool to room temperature and solidify.

4. If you are not allergic to any of the ingredients used in steps 1 and 3, you may test the cooled lip balm on your fingers first, and then apply to your lips. You may take your lip balm home.

**Q1:** How would you rate your lip balm compared to other lip balms you have used?

5. Now you will make four different test products. Begin by labeling four mini foil muffin cups with "10," "20," "30," and "40." (The numbers represent the different concentrations of oil of cinnamon that you will test.) Add 1.0 g (¼ teaspoon) lanolin and 4.0 g (1 teaspoon) beeswax to each muffin cup according to your leader's instructions.

6. Add oil of cinnamon and coconut oil to the muffin cups in the amounts indicated in the following chart.

| Muffin Cup (% Oil of Cinnamon) | Amount of Oil of Cinnamon | Amount of Coconut Oil |
|---|---|---|
| 10 | 1.0 g (¼ teaspoon) | 4.0 g (1 teaspoon) |
| 20 | 2.0 g (½ teaspoon) | 3.0 g (¾ teaspoon) |
| 30 | 3.0 g (¾ teaspoon) | 2.0 g (½ teaspoon) |
| 40 | 4.0 g (1 teaspoon) | 1.0 g (¼ teaspoon) |

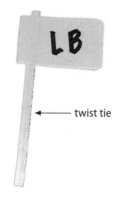

← twist tie

7. Before heating the muffin cups, label one end of a set of paper twist ties the same way you labeled the muffin cups (LB, 10, 20, 30, and 40). Label another set of paper twist ties the same way, but add a W to each label (LBW, 10W, 20W, 30W, and 40W). The W stands for water. Also label two paper twist ties that will not be dipped in any test product with 0 and 0W.

# Camper Notebook

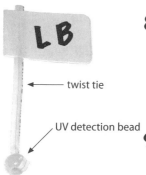

twist tie

UV detection bead

8. Thread a UV detection bead on the unlabeled end of each twist tie. Bend the twist tie's end up to hold the bead on the tie as shown. Repeat the procedure for all the twist ties. These beads will be used in steps 10–14.

**Remember...** Use UV detection beads that turn all the same color.

9. Heat all of the muffin cups (including the LB muffin cup prepared in steps 1–3) on an electric food-warming tray until all ingredients are melted. Stir each mixture occasionally with its own toothpick.

10. Tip the LB muffin cup slightly to maximize the depth of the melted mixture. One at a time, hold the labeled end of the LB and LBW twist ties and completely submerge the beads in the melted mixture. Pull the beads out and allow the coating to harden on the beads (for about 1 minute) before setting the beads down on waxed paper. Repeat this dipping process for the rest of the muffin cups and beads. The O and OW beads will not get dipped.

11. To analyze if your test products are waterproof, place the beads labeled with a W together in a cup of water. The beads should be totally submerged, but the labeled ends should remain out of the water. Let the beads soak for 30–60 minutes.

12. While the beads are soaking in step 11, take the other set of beads outside into direct sunlight to analyze your test products for sunscreen abilities. In the data table at the end of this activity, record the color of each bead. (Beads turn darker as they receive more UV radiation. Lighter beads indicate that the coating blocked the UV radiation. The O bead receives full UV radiation.)

**Q2:** Which beads (besides the O bead) turn the darkest color when exposed to UV light?

**Q3:** Which formulation is best at screening out UV light?

13. After 30–60 minutes, remove the set of beads from the cup of water. Do not dry them off.

**Q4:** Has the test product dissolved off the surface of the beads? How can you tell?

## Camper Notebook

**14.** Repeat step 12 with this second set of beads.

**Q5:** Which of the beads soaked in water (besides the OW bead) turn the darkest color when exposed to UV light?

**Q6:** Which (if any) test product formulation is waterproof? Why or why not?

**Q7:** Would oil of cinnamon make an effective sunscreen? Why or why not?

⚠️ *Oil of cinnamon has not been tested for safety in amounts greater than used as a flavor. Therefore, do not apply any of your test products (those products made after step 4) to skin or lips. Discard all of your test products in the trash. The lip balm containing a couple drops of oil of cinnamon (made in steps 1–3 and labeled as LB) can be taken home and used.*

| Muffin Cup | Oil of Cinnamon | Bead Color After UV Exposure (step 12) | Bead Color After Dipping in Water and UV Exposure (step 14) |
|---|---|---|---|
| LB | 2 drops | | |
| 10 | 10% | | |
| 20 | 20% | | |
| 30 | 30% | | |
| 40 | 40% | | |
| 0 | no product | | |

# Cover Up, Screen, or Block?

## Overview

You've probably heard that using sunscreen and wearing proper clothing and sunglasses will protect you from the sun. In the following activity, you will test the protection of different sunblocks, sunscreens, clothing, and sunglasses using UV detection beads.

## Materials

- ✓ 3 bead setups (prepared by leader)
- ✓ chalk
- ✓ gallon zipper-type plastic bag
- ✓ 4 sun protection products (such as sunscreens and sunblocks) having a wide range of SPF ratings
- ✓ cotton swabs
- ✓ 3 fabric squares with different weaves (prepared by leader)
- ✓ tape
- ✓ clear plastic cup
- ✓ 2 sunglass lenses with different UV ratings (prepared by leader)

## Procedure

### Test of Sun Protection Products

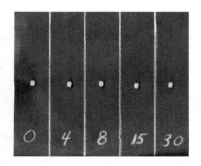

1. As shown at the left, use chalk to label the paper next to each bead in the five-bead setup with the SPF rating of the sun protection product you are going to test. One bead will have no sun protection product (0 SPF) and will be the control. Slide the construction paper into the gallon-sized plastic bag.

2. Use a clean cotton swab for each sun protection product in this step. Smear a small amount of the appropriate product on the plastic bag directly over each bead. It is important to apply the same amount of product evenly over each bead. Use the cotton swab to spread the sun protection product into a circle that is about 1½ inches (about 4 cm) in diameter. Circles of this size should provide protection to the tops and sides of the beads.

3. Write down the name and SPF of each sun protection product and the starting color intensity of each bead in the data table at the end of this activity. Also record the time of day and weather conditions (such as sunny, partly sunny, or cloudy).

Healthy Skin—Cover Up, Screen, or Block?

## Camper Notebook

4. Take the setup outside in the direct sunlight. Without removing the plastic bag, immediately observe and record the color intensity of the beads (such as white, nearly white, light, medium, or dark). If you can't determine the color intensity of the beads through the sunscreens, take the beads indoors, open the plastic bag, and immediately observe the color intensity of the beads (so that the color does not significantly fade).

**Q1:** Compare the color intensity changes of the beads with the SPF ratings of the products covering them. What is the trend?

**Q2:** Imagine doing this test during a commercial to sell sunscreen. Do you think this test would convince people to buy one particular sun protection product over another? Explain your answer.

### Test of Fabric and Sunglasses

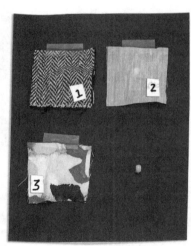

5. Tape each piece of fabric over a bead in a setup as shown at left. Use only one piece of tape for each piece of fabric. The fourth bead will serve as the control. You will leave this bead uncovered to expose it to direct sunlight.

6. With tape, secure the plastic cup and sunglass lenses over the beads in another setup as shown at lower left. The fourth bead will serve as the control.

7. Record the starting color intensity of each bead in the data table. (You can flip up the objects to look at the beads but be sure to re-cover the beads after checking their color intensity.) Record the UV protection rating of the sunglass lenses if you know it. Also record the time of day and weather conditions (such as sunny, partly sunny, or cloudy).

8. Carry the setups outside into the sun. After the control beads turn dark (usually in 1 or 2 minutes), bring the setups indoors and immediately look under the covers to observe the color of the beads. Record the bead color intensity (such as white, nearly white, light, medium, or dark) in the data table.

**Important...** Do not peek at the beads under the objects until you've brought the setups back inside.

# Camper Notebook

**Q3:** Based on your results, what type of clothing do you think would best protect you from the sun? What is your evidence?

**Q4:** Compare the UV protection of the sunglasses to the UV protection of the cup. Why are they different?

| Test of Sun Protection Products | | | |
|---|---|---|---|
| **Name of Product** | **SPF** | **UV Bead Color Intensity** | |
| | | **Start** | **After Sun Exposure** |
| no sunscreen | 0 | | |
| | | | |
| | | | |
| | | | |
| | | | |
| Time of day and weather conditions: | | | |

| | Test of Fabric and Sunglasses | | | |
|---|---|---|---|---|
| | **Type of Cover** | **UV Protection Rating** | **UV Bead Color Intensity** | |
| | | | **Start** | **After 1–2 Minutes** |
| Clothing Test | control (direct sun) | 0 | | |
| | fabric sample 1 | — | | |
| | fabric sample 2 | — | | |
| | fabric sample 3 | — | | |
| Sunglasses Test | control (direct sun) | 0 | | |
| | sunglasses brand 1 | | | |
| | sunglasses brand 2 | | | |
| | cup | — | | |
| Time of day and weather conditions: | | | | |

# Sunning Straws

## Overview

You've probably heard commercials for sunblocks and sunscreens talk about how their products block and absorb ultraviolet (UV) radiation. In this activity, you will actually measure the amount of UV protection that different products offer.

## Materials

- ✓ straws containing a light-detecting chemical (prepared by leader)
  ⚠ *Wear goggles when handling the straws.*
- ✓ permanent marker
- ✓ sunblocks and sunscreens having different sun protection factor (SPF) numbers
- ✓ cotton swabs
- ✓ cardboard or plastic tray to hold the straws horizontally
- ✓ aluminum foil
- ✓ facial tissues or paper towels
- ✓ narrow container (such as a tall cup) to hold the straws upright
- ✓ metric ruler

## Procedure

1. Designate one straw for each sunblock and sunscreen product to be tested by numbering the top of each straw and writing the product names in the data table at the end of this activity. Label one straw "C" for control. This straw will be left as is.

2. Use a cotton swab to spread each sun protection product on the entire outside surface of the appropriate straw. Be sure to use a different cotton swab for each product. Place the straws horizontally on a tray covered with aluminum foil, leaving as much space between the straws as you can.

3. Take the tray outside and expose the straws to direct sunlight for at least 90 minutes.

4. After exposing the straws to the sun, bring the straws inside and wipe the sun protection products off the outside using facial tissues or paper towels.

## Camper Notebook

**Q1:** What do you see in the straw that had nothing applied on the outside (the control)?

**5.** Look at each straw. If any solid is present in the straw, hold it upright and gently tap the side of the straw to allow the solid to settle to the bottom of the straw. The solid is called benzopinacol. Store the straws upright in the narrow container.

**6.** After allowing the solid to settle for a few minutes, use a metric ruler to measure the height of the white product present in each straw. Record the heights in the data table.

**Q2:** How is this procedure a quantitative measure of the amount of UV radiation hitting the solution inside each straw?

**Q3:** Which product offers the best protection against UV radiation?

|   | Sunblock or Sunscreen Product | Height of White Solid |
|---|---|---|
| C | no product (control) |   |
| 1 |   |   |
| 2 |   |   |
| 3 |   |   |
| 4 |   |   |
| 5 |   |   |
| 6 |   |   |
| 7 |   |   |

# Suntan in a Bottle

## Overview

Sunless tanning products are a popular way for people to appear tan without exposing themselves to harmful UV rays. In this activity, you'll test a sunless tanning product and observe how it affects different kinds of fabric.

## Materials

- ✓ newspaper
- ✓ different types of white fabric (such as cotton, wool, polyester, nylon, and silk)
- ✓ scissors
- ✓ glue
- ✓ sunless tanning product containing dihydroxyacetone (DHA)
- ☞ *Read the safety precautions on the product label. Be careful not to get the product on your clothes. Wash your hands thoroughly after applying the product.*
- ✓ cotton swab

## Procedure

1. Lay down newspaper to protect the work area. Cut two squares from each fabric, each measuring about 1½ inches × 1½ inches (about 4 cm × 4 cm).

2. Glue the two squares of each fabric type in the data table at the end of this activity. (One square will be untreated, serving as the control, and the other will be treated with a sunless tanning product.) In the data table, record the types of fabric. (It's okay if you don't know that information.) Let the glue dry.

3. Carefully use a cotton swab to evenly apply sunless tanning product to each square of fabric in the "Treated Sample" column of the data table. Notice what time it is. Thoroughly wash your hands when you are done.

**Q1:** What is the color of the sunless tanning product when it is first applied?

## Camper Notebook

**4.** Observe the results after several hours. Record the total treatment time and your observations in the data table.

**Q2:** Which fabrics react with the sunless tanning product and become "tanned"?

**Q3:** What do the "tanned" fabrics have in common? (Hint: Think about what each fabric is made of.)

## Camper Notebook

| Fabric Treated for _____ Hours and _____ Minutes |||| 
|---|---|---|---|
| Type of Fabric | Untreated Sample (Control) | Treated Sample | Observations |
| | Glue Fabric Here | Glue Fabric Here | |
| | Glue Fabric Here | Glue Fabric Here | |
| | Glue Fabric Here | Glue Fabric Here | |
| | Glue Fabric Here | Glue Fabric Here | |
| | Glue Fabric Here | Glue Fabric Here | |

# A Look at Bleaching

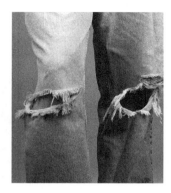

## Overview

Some people think they look better if they are suntanned, while other people want their skin to be lighter. Either tanning or skin bleaching to change skin color is risky. In this activity, you'll get an idea of the risks of skin bleaching by observing the effects of different household products on fabric samples. (Skin testing will not be done for safety reasons.)

## Materials

- ✓ large plastic garbage bag
- ✓ tea-dyed fabric (prepared by leader)
- ✓ several dark-colored fabric samples (such as black cotton broadcloth, black or blue cotton denim, black wool, black nylon, and black polyester)
- ✓ scissors capable of cutting fabric
- ✓ ruler
- ✓ rubber gloves
- ✓ up to 6 household and skin care products such as:
    - liquid or gel dishwasher detergent
    - laundry bleach pen
    - anti-aging cream containing hydroquinone
    - acne care product containing benzoyl peroxide
    - "oxygen" cleaner (such as Oxiclean™ or OxiMagic™)
    - tooth whitening gel
    - hydrogen peroxide

  ⚠ *Do not use undiluted liquid household laundry bleach in this activity.*
- ✓ applicators for each product such as:
    - cotton swab (such as Q-Tip®)
    - small sponge piece no larger than 1 inch × 1 inch (2 cm × 2 cm)
- ✓ hand soap

## Procedure

1. Cover your work area with a large plastic bag to prevent damage to the surface if test substances soak through the fabrics.

## Camper Notebook

top of fabric

2. If not done already, cut each fabric sample into a large rectangle about 6 inch × 8 inch (15 cm × 20 cm). Mark one edge of each piece as the top by cutting off both corners as shown in the figure at left. (You'll need to know which edge is the top in order to keep track of where you've placed the test products on the fabrics.)

3. Use the figure at left to record where you will place each product to be tested. Be sure to evenly distribute the locations, allowing at least a 2-inch × 2-inch (5-cm × 5-cm) area for each product. Record the name of each product and its bleaching agent on the figure.

☞ *Read the product labels or ask the leader to identify the bleaching agents. Examples include sodium hypochlorite, hydroquinone, benzoyl peroxide, sodium percarbonate, and hydrogen peroxide.*

4. Put on rubber gloves. Use a swab or sponge applicator to apply each product to each fabric. A mark no bigger than about 1 inch × 1 inch (1.5 cm × 1.5 cm) works well because some products will spread out on the fabric over time. Be sure to use a different applicator for each product.

☞ *Follow any safety precautions listed on the label when handling each product.*

5. Allow the products to remain on the fabrics for at least 30 minutes. When time is up, put on rubber gloves and rinse the fabrics. Then, wash the fabrics with hand soap to remove any oily residue. Squeeze out as much water as possible. Roll the fabric samples in paper towels and twist the paper towels to remove excess moisture.

6. Record each type of fabric, each test product, and your observations in the data table at the end of this activity.

**Q1:** Generally, what do the household products do to the fabrics?


**Q2:** Which fabrics are most sensitive to the bleaching agents in the different products?

# Camper Notebook

| Test Product | Tea-Dyed | | | | | |
|---|---|---|---|---|---|---|
| | | | | | | |
| | | | | | | |
| | | | | | | |
| | | | | | | |
| | | | | | | |
| | | | | | | |
| | | | | | | |

Observations for Each Type of Fabric

Healthy Skin—A Look at Bleaching

# UV Detective Challenge

Does being outdoors in the shade of a tree provide protection from UV radiation compared to direct sunlight? Let's find out!

## What You'll Need:

- at least 5 UV detection beads in assorted colors
  - *UV beads are available from science and teacher supply stores or similar sources on the Internet. UV beads in assorted colors are preferred.*
- Any one of the following strap materials:
  - pipe cleaners (chenille stems)
  - leather or leatherlike "rawhide" lacing
  - nylon cord, narrow ribbon, or yarn
  - narrow elastic cord (beading elastic)
- twist ties or something else to label the beads
- container that doesn't let in any light (such as a dark bag, dark box, or your pocket)
- (optional) colorless fingernail polish

## What You'll Do for Part 1:

>> 1. Cut the strap material so it fits around the wrist. String the UV detection beads on the strap. Tie or twist the strap to make a bracelet. (If using elastic for the strap, put some colorless fingernail polish onto the knot after it is tied to help prevent the knot from fraying and coming undone.)

>> 2. While indoors, decide and mark which beads on your bracelet will be numbers 1 and 2. (See photo.) The other bead numbers will consecutively follow along the strap. In the Part 1 data table at the end of this activity, record the starting color of each bead (such as white, nearly white, light, medium, or dark).

This is Bead 2.
This is Bead 1.

>> 3. Put the bracelet in a dark container. Bring the bracelet in its container outside to a shady area (such as the shade of a building, tree, or umbrella).

>> 4. Remove the bracelet from its dark container, wait a minute or two, and then observe the color changes of the beads. Record your location and the shade each bead turns.

**Take-Home Activity: UV Detective Challenge**

© 2007 Terrific Science Press™
The publisher takes no responsibility for use of any materials or methods described in this monograph, nor for the products thereof. This publication was made possible by Grant Number 1 R25 RR16301-01A1 from the National Center for Research Resources (NCRR), a component of the National Institutes of Health (NIH). Its contents are solely the responsibility of the authors and do not necessarily represent the official views of NCRR or NIH.

**» 5.** Put the bracelet back in the container until the colors fade and then repeat step 3 in two other shady locations.

**» 6.** For comparison, bring your UV detecting bracelet into direct sunlight. Observe and record the shades of the beads as accurately as possible.

## What You'll Do for Part 2:

**» 1.** Use the procedure in Part 1 to test all or some of the following materials and locations to see if UV radiation is blocked. Create a data table and record your results.
- car windows such as front windshield, side windows, and back window
- your sunglasses, someone else's sunglasses, or regular eyeglasses
- house windows
- tanning bed
- underwater at the pool (try different water depths) or in a bucket of water

**» 2.** Think of other places to test with your bracelet and try them. For example, how long after sunrise or before sunset does your bracelet detect UV light?

## Questions to Consider:

- Does shade provide some protection from UV radiation?
- Why did you carry the bracelet to the shade in a container that doesn't let in any light?
- Why did you observe the color that the bracelet turned in direct sunlight after you observed the color that the bracelet turned in the shade?
- Is there a difference in the amount of detectable UV radiation in different areas of shade? For example, are the beads colored differently in the shade from a building or the shade from a tree? If you go deeper into natural shade, does the color of the beads change? If you are in the shade of an umbrella, how much UV light is detected?
- If your bracelet was made up of different colored UV beads, does each color respond to UV light to the same extent? Why or why not?

## What's the Deal?

Shade created by a building or tree is probably less protective than clothing and sunglasses. Even in this shade, reflected UV can get to the beads. If an object is either directly or indirectly illuminated by sunlight, it is receiving at least some UV. Going deeper into shade will reduce but not eliminate UV exposure. People on beaches and boats often get a tan even if they are in the shade because of UV reflection off the sand and water.

By placing the bracelet in the dark before bringing it out in the shade, the beads can change more drastically from colorless to colored. It is easier to detect this color change rather than to wait for the bright colors to fade to constant lighter colors. It is easier to observe the brightest color after observing a faint color, for a similar reason. Each dye in the different-colored UV detection beads is a different chemical compound with a different response to UV light, so each bead color has its unique sensitivity to UV radiation.

| UV Bead Shade | | | | | |
|---|---|---|---|---|---|
| Bead | Start | Location: _____ | Location: _____ | Location: _____ | Direct Sunlight |
| 1 | | | | | |
| 2 | | | | | |
| 3 | | | | | |
| 4 | | | | | |
| 5 | | | | | |
| 6 | | | | | |
| 7 | | | | | |
| 8 | | | | | |

# LEADER GUIDE

These notes will help you plan and execute the hands-on activities in your camp.

LEADER

**LEADER**

# TOPIC 1
## DISEASE CONTROL IS IN YOUR HANDS

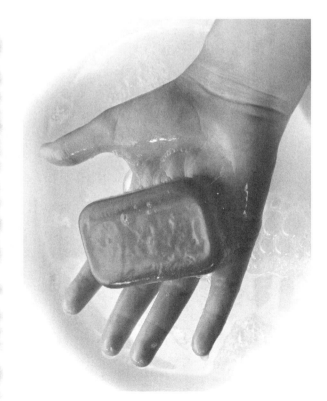

**List of Activities**

- ✓ Why Soap? ..................................................136
  *Try different hand-washing methods to learn which one works best at removing germs.*

- ✓ Design Your Own Soap ..................................140
  *Choose additives such as color, fragrance, and decorative objects to design your own soap.*

- ✓ DNA from Strawberries .................................143
  *Observe how soaps and detergents affect the oil-like cell walls of strawberry cells.*

- ✓ Colorful Lather Printing ................................147
  *Learn about the concepts of polarity and hydrophobicity while making fun marbling patterns on paper.*

- ✓ Surfactants .................................................150
  *Learn how soaps and detergents work to remove dirt and germs from the skin.*

- ✓ Blowing Bubbles ..........................................154
  *Play with bubbles and learn details about their structure.*

- ✓ Take-Home Activity: Watching Granny Smith Rot ......158
  *Discover the importance of hand washing before handling food.*

- ✓ Take-Home Activity: How Clean Is Your Clan? ..........159
  *Observe how well your friends and family wash their hands.*

Page 135

LEADER

# LEADER

## Leader Guide

## Why Soap?

> **Suggestions for leading this activity begin on the next page.**

According to the Centers for Disease Control (CDC), the most effective way to slow or stop the spread of disease is washing hands with soap and water. By giving campers a simulated infection, you will be able to demonstrate how quickly and easily infections are passed from person to person by casual contact. Campers then investigate and rank the effectiveness of some hand washing methods in removing germs from the hands.

## Some Leader Background Information

Disease microbes pass from person to person in many different ways. One important method is through hand contact. For example, pink eye (conjunctivitis) is a common eye infection that can be caused by either bacteria or viruses. If you have pink eye and you rub your eye and touch a doorknob, you can pass the infection to someone else who touches the doorknob. Another common way microbes are spread is by sneezing. When you sneeze, you release a cloud of tiny droplets. If you have a cold, each droplet will contain some cold viruses. Another person breathing in the droplets can then catch the cold. Other ways that microbes can get into your body are through your mouth, insect bites, and cuts in the skin.

In this activity, the ultra fine glitter serves as a visual reminder of the germs and other microorganisms (microbes) found on our hands, although the glitter pieces are much bigger than actual microbes. While germs can be harmful because they can cause disease or infection, other microbes are necessary to keep us healthy.

Step 2 results will vary based on the amount of glitter mixture applied, amount of contact with the "infected" hand or object, condition of the hands, and other variables. Typically, in the hand-shaking simulation, the glitter will be passed at least halfway down the line. Campers may be surprised that traces of the glitter mixture are carried as far as they are.

Step 6 results probably revealed that water, soap, and vigorous scrubbing for 20 seconds will result in the cleanest hands. Either just wiping with a paper towel or just rinsing with plain water is the least effective method. You may have noticed that wiping with a paper towel will remove some of the glitter, while air drying will not. Some studies indicate that drying with a clean paper towel is slightly more effective for removing germs from the hands than drying with a hand dryer.

# Why Soap?

**Leader Guide**

## Materials for Camper Procedure—See Left

## Materials for Getting Ready

- ✓ petroleum jelly (such as Vaseline®)
- ✓ ultra fine cosmetic-grade glitter
- ☞ Dark-colored glitter (such as red or green) works best. Do not use craft glitter.
- ✓ zipper-type plastic bag
- ✓ measuring spoons
- ✓ scissors
- ✓ (optional) Glo Germ™ gel and ultraviolet (UV) light
- ☞ Glo Germ is available at www.glogerm.com.

## Getting Ready

1. Prepare a glitter mixture in a zipper-type plastic bag by adding two parts petroleum jelly and one part ultra fine glitter. Gently knead the bag to mix the glitter into the petroleum jelly. To dispense the mixture, cut a corner out of the bag as shown. A pea-sized portion works well for adult-sized hands.

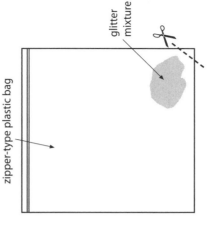

zipper-type plastic bag · glitter mixture

2. Decide how you will expose campers to a simulated infection for the group introduction in step 1 of the procedure. Here are some ideas:

- Have some, but not all, of the camp leaders apply glitter mixture to their hands before campers arrive to simulate the presence of microbes. Rub the mixture thoroughly into the hands until they don't feel greasy anymore. Shake hands with campers as they arrive. Some campers will be infected and others will not.

Page 137

Disease Control Is in Your Hands—Why Soap?

**LEADER**

---

# Why Soap?

### Overview

How do disease-causing germs get on your hands and how can you get them off? In this activity, you will learn how diseases can spread. You will also experiment with different hand-washing methods.

### Materials

- ✓ glitter mixture to represent infection (provided by leader)
- ✓ paper towels
- ✓ access to sink with warm water or pitchers of warm water plus a bucket for waste water
- ✓ hand soap
- ✓ stopwatch or timepiece with a second hand
- ✓ hair dryer

### Procedure

1. Participate in a group introduction as directed by your leader.
2. Examine everyone's hands to discover who has been "infected." You may need to use a special light.

Q1: How did the infection get passed around? How many people became infected?

3. Work with a friend to test the various methods of removing germs from hands listed in the data table at the end of this activity. One person should be the hand washer and the other person should be the timekeeper and data recorder.
4. Squeeze a pea-sized portion of glitter mixture in the hand of the hand washer and have her or him rub the hands together so the mixture thoroughly covers the front and back of each hand. The glitter mixture represents germs.
5. Have the hand washer try the first method listed in the data table. Then, the data recorder should shade the hand diagrams to show where glitter and oily sheen remains on the palm and back of each hand.

Page 34

Disease Control Is in Your Hands—Why Soap?

**CAMPER**

# Why Soap?

## Leader Guide

- Divide campers into groups of 6 or 7. Ask each group to form a line, explaining that the first person in line will act as the germ carrier who is presently infected with a cold or disease. Have the first person in line thoroughly apply glitter mixture to the palms and backs of their hands, then firmly shake the hand of the next person in line. Continue shaking hands down the line until the end.

- Select objects that can be easily washed or discarded (such as pencils, pens, and plastic silverware). Apply glitter mixture to some of the objects. Devise a reason for campers to handle the objects (you choose whether or not to tell them about the "infection"). For example, campers can use treated pens to sign in when they arrive to do the activity or campers can just pass the objects around the room. Putting glitter mixture on objects that already have glitter in their design helps with the disguise.

- As an alternative, Glo Germ gel can be used instead of glitter mixture with any of the above ideas. Since Glo Germ can only be seen under a UV light, using this gel will add an element of surprise. Practice with Glo Germ ahead of time to determine the best amount of gel to apply. Use a UV light in step 2 of the procedure to reveal how the gel spread during the simulated infection.

## Procedure Notes and Tips

- If only some campers are "infected" (without their knowledge) in step 1, waiting an hour or two before revealing the presence of the infection will show how it spread over time from camper to camper through direct contact or through the use of objects in the room. After the infection is revealed, campers may enjoy figuring out exactly how the infection spread.

- If the infection is Glo Germ rather than glitter, the infection can be detected in step 2 as fluorescence under a UV light.

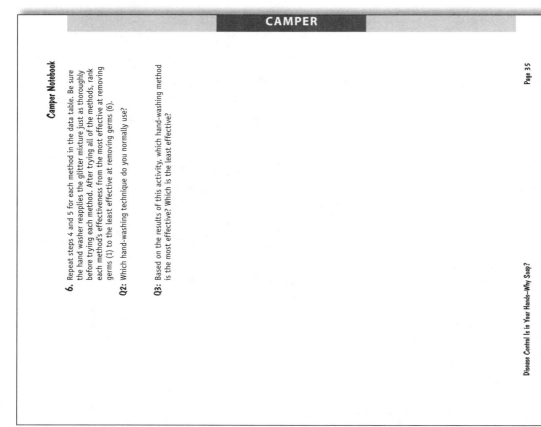

### Camper Notebook

6. Repeat steps 4 and 5 for each method in the data table. Be sure the hand washer reapplies the glitter mixture just as thoroughly before trying each method. After trying all of the methods, rank each method's effectiveness from the most effective at removing germs (1) to the least effective at removing germs (6).

Q2: Which hand-washing technique do you normally use?

Q3: Based on the results of this activity, which hand-washing method is the most effective? Which is the least effective?

# Why Soap?

## Leader Guide

**Q1:** Answers will vary depending on the method chosen and other conditions (such as dryness of the hands and firmness of the handshakes).

**Q2:** Answers will vary, but one study of middle and high school students showed that female students used soap only 28% of the time when they washed their hands. Male students, in contrast, used soap only 8% of the time.

**Q3:** Answers will vary, but water, soap, and vigorous scrubbing for 20 seconds usually results in the cleanest hands. Either just wiping with a paper towel or just rinsing with plain water is the least effective method.

## Wrap-Up

Review with campers the "How to Wash" and "When to Wash" lists below. Point out that the most important part of good hand washing is scrubbing with soap for at least 20 seconds (while your hands are not under running water). Emphasize the importance of hand washing at the appropriate times throughout the day. At the end of the camp day, look again for the presence of "infection" on the campers' hands. While some may still be present (probably around and under the fingernails), the amount will be quite small compared to the first time they checked. Explain to the campers that this residual amount of glitter mixture shows that not all germs can be eliminated. But the fewer germs there are to potentially infect someone, the less likely that person will get sick.

### How to wash...

1. Wet hands under warm running water and apply soap.
2. Keeping hands away from the running faucet, rub hands together swiftly for at least 20 seconds. The soap should be bubbly.
3. Be sure to wash all surfaces of the hands, including between the fingers and under fingernails. Don't forget to include wrists.
4. Rinse well until all soap is gone.
5. Dry hands with a clean towel.

### When to wash...

- before, during, and after preparing food
- before eating
- before and after treating a cut or wound
- after blowing your nose, coughing, or sneezing
- after using the bathroom
- after handling animals or animal wastes
- after using household chemicals
- after changing a diaper
- when hands are visibly dirty
- more frequently when someone in your home is sick

## Camper Notebook

| Method | Results | Observations |
|---|---|---|
| Wipe with a dry paper towel for 5 seconds (don't use soap and water). | palm / back | |
| Rinse with plain water for 5 seconds (but don't dry hands). | palm / back | |
| Rinse with plain water for 5 seconds and dry with a paper towel. | palm / back | |
| Wash with soap for 5 seconds, rinse with water, and dry with a paper towel. | | |
| Wash with soap for 5 seconds, rinse with water, and dry with a hair dryer. | | |
| Thoroughly wash with soap for 20 seconds while hands are not under faucet, rinse with water, and dry with a paper towel. | | |

## LEADER

# Design Your Own Soap

## Leader Guide

> ✋ **Suggestions for leading this activity begin on the next page.**

After campers learn what ingredients make up soap, they make their own soap with glycerin soap base, color, and other additives. This activity is popular, quick, and gives campers something to take home at the end of the day.

## Some Leader Background Information

Chances are that many of the skin cleaning items in your shower and on your sink aren't soap at all, but synthetic detergents filled with ingredients designed not only to clean your body, but to deodorize, sanitize, and moisturize. Next time you're at the store, take a close look at the products in the "soap" aisle. You'll discover that some popular brands don't even have the word soap on the label; instead they advertise themselves as "bath" or "beauty" bars.

Soap is a compound made by reacting fat or oil with a base (usually sodium or potassium hydroxide). The reason why most people don't use soap anymore is simple: the detergent products work better. They make suds more easily in water and don't form gummy deposits (soap scum) in your bathtub.

Soap has been around for at least 2,300 years. At first, soap was used only as an ointment and medication. Its value as a skin cleaner wasn't recognized until the 2nd century A.D. The ancient Celts called it saipo, from which we get our modern word "soap."

In the old days, people made soap from animal fat and wood ashes. These two materials were boiled in water to allow a reaction called saponification to occur. This process is chemically very similar to the one used to make soap today. Animal fat (called tallow) is still the most common source of fat used, but other sources of fats such as palm oil, coconut oil, and olive oil are also used. Wood ashes (the base) have been replaced by sodium hydroxide (lye) or potassium hydroxide (potash).

In addition to producing soap, the saponification process produces glycerin as a second product (called a by-product). In most cases, glycerin is removed in the purification process and sold separately. However, some soap makers keep the glycerin in their products and some even add more. Glycerin provides a moisturizing effect to the product as well as giving it a translucent appearance.

The first synthetic detergents for general use were made in Germany during World War I. Most detergents are made by reacting certain chemicals with petroleum by-products. Though they were invented during wartime to save on the use of fats and oils, detergents have since developed into sophisticated products that in many ways are superior to soaps.

Today, soaps and detergents come in powdered, liquid, foam, and solid form. In public restrooms, foam or liquid hand cleaner is preferred to bar cleaner because most people don't like to pick up a bar covered with someone else's leftover dirt.

# Design Your Own Soap

## Leader Guide

### Materials for Camper Procedure—See Left

### Supply Information

- Soapmaking supplies are available from craft and hobby stores.

### Getting Ready

1. Depending on the ability of the campers and the available camp time, you may want to cut the glycerin soap base into small pieces ahead of the camp session.

2. Decide which method campers will use to melt the glycerin soap.

3. Gather a selection of cosmetic-grade soap additives for campers to choose from in step 3 of the procedure. Here are some examples:

| Additive | Examples |
| --- | --- |
| scent | soapmaking fragrance, body mist, or perfume |
| color | food color or soapmaking colorants |
| oil | coconut, sweet almond, refined olive, avocado, jojoba, or lanolin |
| solid | dried flowers and herbs, grated citrus peel, ultra fine cosmetic grade glitter, or small plastic toys |

**To make fun fish soap...**
Pour a layer of melted soap into a plastic bag, add a toy fish, and pour another layer of soap on top.

Page 141

---

## CAMPER

### Design Your Own Soap

#### Overview

Soap casting is currently a popular craft. Have you ever tried it? In this activity, you get to design your own soap by choosing additives such as color, fragrance, and decorative objects.

#### Materials

✓ about ½ cup (125 mL) glycerin soap base (such as Neutrogena® or another, less expensive brand)
✓ knife (use with adult supervision)
✓ cutting board
✓ one of the following sets of materials to melt the glycerin soap:
  • microwave-safe container, plastic food wrap, and microwave
  • pan and hot plate or stove
✓ wooden chopstick or wooden spoon
✓ cosmetic-grade soap additives (such as scent, color, oils, and solids)
✓ flexible mold (such as soapmaking mold, disposable paper or plastic cup, or small plastic bag)
✓ materials to measure additives (such as droppers, measuring spoons, and cups)
✓ (optional) 70% isopropyl rubbing alcohol in spray bottle
✓ (optional) access to a refrigerator

#### Procedure

1. With adult supervision, carefully cut approximately ½ cup (125 mL) glycerin soap base into small pieces measuring no bigger than 1 inch x 1 inch (3 cm x 3 cm).

2. Follow one of these two methods to melt the glycerin soap pieces:
   • Microwave method: Place the glycerin pieces in a microwave-safe container. Cover the container with plastic wrap (to prevent loss of moisture) and microwave in 10-second intervals, stirring in between with a wooden chopstick or wooden spoon. (Remove from the microwave after melted, usually after about 30 seconds.)

Page 37

# LEADER

## Design Your Own Soap — Leader Guide

4. Read the Cosmetic-Grade Soap Additives table in the Camper Notebook to familiarize yourself with the specific instructions for adding additives to the glycerin.

## Introducing the Activity

Tell campers what soap used to be made of and what it is made of today. (For help with the discussion, see Leader Background Information.) Show them the materials they can use to make their own soap.

---

# CAMPER

## Camper Notebook

- Stove top or hot plate method: Place the glycerin pieces in a pan. Carefully heat the pan while stirring with a wooden chopstick or wooden spoon. Remove the pan from the heat when all of the glycerin soap is melted.

3. Decide on the additives you want to use in your soap. Use the wooden chopstick or spoon to stir the additives into the melted soap according to the directions provided in the table below.

4. Pour the melted glycerin-additives mixture into the flexible mold.
   *If bubbles form on the surface of the melted glycerin, you can lightly spray the surface with 70% isopropyl rubbing alcohol.*

5. Allow the soap to cool to room temperature. (This may take a couple of hours.) Placing the mold in the refrigerator or freezer will speed up the cooling process, but cooling the soap much past room temperature may make it too hard and flaky.

6. Tap and flex the mold to remove the hardened soap. Bag and cup molds can be cut away or torn. Store your soap in plastic food wrap to prevent loss of moisture until you are ready to use it.

### Cosmetic-Grade Soap Additives

| Additive | Examples | Instructions |
|---|---|---|
| scent | soapmaking fragrance | Add about ¼ teaspoon (1.25 mL) commercial soapmaking fragrance. With milder scents, add more fragrance with a dropper until the desired scent is achieved. |
|  | body mist | Add 1 tablespoon (15 mL) body mist to the melted glycerin. Adjust the quantity until the desired scent is achieved. |
|  | perfume | Allow the melted glycerin to partially cool before adding alcohol-based perfumes because the fragrance will evaporate quickly if the glycerin is too hot. |
| color | food color | Add a drop or two of food color. (The color won't come off on hands or towels when the soap is used.) |
|  | soapmaking colorants | Follow instructions on the package. |
| oil | coconut, sweet almond, refined olive, avocado, jojoba, or lanolin | Add two or three drops of oil. |
| solid | dried flowers and herbs, grated citrus peel, ultra fine cosmetic grade glitter, or small plastic toys | Pour a layer of glycerin into the mold, letting the glycerin begin to harden. Add the solids, and pour another layer of glycerin on top. Fresh flowers and other plant materials may require preservative to prevent mold growth in the soap. When placed in soap, some fresh flowers such as rose petals and lavender may turn brown with age. |

## Leader Guide

# DNA from Strawberries

> Suggestions for leading this activity begin on the next page.

Microbes that cause disease are passed from person to person in many ways, including hand contact. Soaps and detergents are essential in removing germs from hands because germs (like dirt) are often attached to hands with oil or grease. Cleaners attract and surround oil and grease, allowing germs and dirt to be lifted from the surface being cleaned and then rinsed away. Soaps and detergents also have a similar effect on the "oil-like" walls and outer membranes of cells. In this activity, soaps and detergents are used to partially disrupt and break open plant cell walls, allowing DNA from inside the cells to be collected. By doing this, campers get to see a secondary way that soaps and detergents can reduce some transmittable germs, since germs with broken cell walls or membranes are rarely viable. Since genomic research begins with DNA, campers are also exposed to the first step in genomic research.

## Some Leader Background Information

A common procedure for the isolation of DNA (the genetic material inside a cell) calls for the use of a detergent (or surfactant) and salts. Some procedures use liquid dish detergent or shampoo. The detergent's "oil-like" part disrupts part of the oily (or waxy) cell membrane or cell wall, causing the membrane or wall to be opened. (Depending on the type of cell, the outside of a cell is either a wall or membrane. Plant cells, such as strawberries, have cell walls.) Once the cell is open, the DNA inside the cell can be isolated. Salt is added to increase the solubility of proteins in the water layer and to disable enzymes that may otherwise destroy the cell's DNA as the cell is opened. Some enzymes destroy DNA (if the cell is damaged) to prevent changes in the genetic code of the DNA being passed on. Salt prevents the enzymes from destroying the DNA, thereby allowing the DNA to be isolated.

Isolation of DNA is interesting for two reasons. First, conditions that allow the strawberry cell wall to be disrupted might be considered biocidal. (Since the strawberry cell's exposure to the detergent in this procedure is longer and different from hand bacteria exposure to a detergent in actual hand washing, it may not be correct to consider soaps and detergents themselves antibacterial). Secondly, DNA isolation from strawberries is interesting as the first step in revealing and understanding the genetic code of strawberries.

In this activity, strawberry cells are treated with a surfactant and salt. After opening the cells, DNA from inside the cell is isolated as milky strands on a wooden skewer in the presence of 99% isopropyl alcohol. (DNA is not soluble in 99% isopropyl alcohol.) Campers discover that DNA cannot be isolated if the cleaner is left out of the procedure. Different surfactants (detergent, shampoo, and soap) are studied by different campers and the amount of collected DNA is compared.

The strawberry cell walls contain an oily or waxy surface (made of ordered hydrophobic molecules) that holds the water and interior cell parts inside the cell. The soap or detergent acts on the oil or waxy cell wall similar to the way soap or detergent acts on oil or greasy dirt on skin or fabric. The action of soap or detergent on greasy or oily dirt is described in the "Why Soap?" and "Surfactants" activities in this collection of activities on soaps and detergents.

This activity is adapted from Sweeney, D. "Berry Full of DNA," *Biology: Exploring Life;* Pearson Education, 2001.

# LEADER

## DNA from Strawberries — Leader Guide

### Materials for Camper Procedure—See Left

### Materials for Getting Ready

✓ ice and ice bath
✓ bottle of 99% isopropyl alcohol

### Getting Ready

1. Prepare an ice bath.
2. Place the bottle of isopropyl alcohol in the ice bath so that its contents are ice cold when the campers use it.

### Introducing the Activity

Describe how the hydrophobic part of soaps and detergents dissolves some of the greasy part of dirt, allowing dirt and germs to disperse into water and be washed away. (For help with the discussion, see Leader Background Information in the "Why Soap?" and "Surfactants" activities.) Tell campers that they will observe the effects of soaps and detergents on the oily (waxy) surface of plant cells.

---

## CAMPER

## DNA from Strawberries

### Overview

Today, we know that many diseases are spread by germs people carry on their hands. We know that soap is important when washing our hands, but how does soap work to clean our hands? In this activity, strawberry cells with oil-like cell walls are related to oily and greasy dirt. You will discover what different soaps and detergents do to the strawberry cell walls.

### Materials

✓ 2 small zipper-type plastic bags (Freezer bags are preferred to sandwich bags because they are thicker. Do not use bags with bottom pleats.)
✓ 2 ripe strawberries
✓ 6 pinches table salt (sodium chloride, NaCl)
✓ 3 small cups (such as disposable bathroom cups or measuring cups that come with liquid medicines)
✓ water
✓ teaspoon or 10-mL graduated cylinder
✓ soap- or detergent-based cleaner, such as Ivory® bar soap, liquid dishwashing detergent, or shampoo without conditioner
✓ laboratory spatula or butter knife and measuring spoon (if using Ivory bar soap)
✓ dropper (if using liquid dishwashing detergent or shampoo)
✓ 4 test tubes
✓ 60° angle funnel
✓ 2 filter papers or cone-type coffee filters
✓ ice bath
✓ ice-cold 99% isopropyl alcohol (prepared by leader)
✓ 2 wooden skewers

### Procedure

1. Place one strawberry in each plastic bag. Seal the tops of the bags and knead the bags to completely mash the strawberries. Add 3 pinches table salt to each bag and mix again.

# DNA from Strawberries

## Leader Guide

### Procedure Notes and Tips

- It is easier to obtain the DNA from ripe strawberries rather than unripe ones. Ripe berries contain pectinases and cellulases that are already partially breaking down the cell walls as the strawberry ripens and becomes juicy.
- Have different campers test different cleaners. They will share and compare their results with other campers in the Wrap-Up.
- DNA is insoluble in isopropyl alcohol. This activity uses 99% isopropyl alcohol because, the less water the isopropyl alcohol contains, the more DNA will be isolated.

---

## Camper Notebook

2. Your leader will tell you which cleaner solution to make. Prepare the cleaner solution by putting 2 teaspoons (10 mL) water into a small cup, then doing one of the following:
   - For a soap solution, use a laboratory spatula or butter knife to scrape less than ¼ teaspoon (1.25 mL) Ivory bar soap into the cup. Stir until the soap is dissolved.
   - For a dishwashing detergent solution, add 5 drops dish detergent into the cup and swirl.
   - For a shampoo solution, add 10 drops shampoo into the cup and swirl.

3. Add all of the cleaner solution you made in step 2 into one bag containing a mashed strawberry and label the bag with the type of cleaner solution you prepared. Add 2 teaspoons (10 mL) water to the other bag containing a mashed strawberry and label the bag "no cleaner." (See left.) Gently knead each bag for about 2 minutes.

Your label may differ.

4. Put aside the bags of strawberries for a moment. Label two test tubes like the bags in step 3 (one with the type of cleaner solution you prepared and one with "no cleaner"). Add ice-cold isopropyl alcohol provided by your leader to each test tube to a height of about 2 inches. Store the test tubes in an ice bath. (See right.) These test tubes will be used to store the wooden skewers at the end of step 7.

Setup described in step 4 for storage of DNA collected in steps 7 and 8

5. Label two clean small cups like the bags in step 3. Pour each strawberry mixture through a funnel lined with filter paper and into the appropriate cup. (See left.) Be sure to clean the funnel and use new paper after filtering the first mixture. Discard the bags and filters containing the strawberry solids into the trash.

6. Label two clean test tubes like the bags in step 3. Pour the strawberry cell extracts from the cups into the corresponding test tubes to a height of about 1 inch. Place the test tubes in the ice bath.

# LEADER

## DNA from Strawberries

### Leader Guide

**Q1:** No milky fibers (DNA) are collected from the "no cleaner" test tube. Milky fibers are collected from the test tube with the cleaner, indicating that the cleaner causes the cell walls to break open.

**Q2:** No DNA is isolated when the cleaner is left out because the cell walls are not broken.

### Wrap-Up

Ask campers to observe and compare the skewers of the other campers. They should notice that all of the "no cleaner" skewers hold no DNA, but the different cleaner skewers have varying amounts of DNA. Discuss the results of the activity with the campers. In general, detergents have been preferred to soaps at disrupting the lipid (fat or oil) material around the outside of the cell. Some detergents are better than others. Point out that some disease-causing bacteria can be opened with detergent similar to what was just accomplished with strawberry cells, but not all bacterial cells have the same outside structures as strawberry cells. Even if soaps and detergent don't break down bacteria cells, they still break down greasy dirt so dirt and germs can be lifted from the surface being cleaned and then rinsed away.

---

## CAMPER

### Camper Notebook

**Tip...** It may be necessary to return the test tube to the ice bath periodically during step 7. Since DNA is less soluble in cold solutions, you will be able to collect more DNA when the test tube is kept cold.

7. Carefully pour about an inch of ice-cold isopropyl alcohol provided by your leader on top of the cold, cleaner-treated extract. Do not mix or swirl. Observe the interface between the strawberry extract and isopropyl alcohol. If milky strands (DNA) form at the interface, use a twisting motion to wind the milky fibers onto a wooden skewer. Continue twisting until no more DNA can be wound on the wooden skewer. Store the skewer in the appropriately labeled test tube you prepared in step 4 that contains just isopropyl alcohol. Keep the test tube in the ice bath.

8. Repeat step 7 with the cold strawberry extract not treated with cleaner.

9. Briefly remove both skewers from the isopropyl alcohol and compare the results of the two extractions. Return the skewers to the test tubes to save them for the Wrap-Up.

**Q1:** How do the two skewers differ? Explain the results.

**Q2:** Strawberry DNA is genetic material located inside the cell walls of strawberry cells. In step 8, how much DNA did you isolate from the cold strawberry extract not treated with cleaner? Why?

10. To clean up, throw away the skewers and DNA in the trash, rinse the strawberry extracts and other liquids down the drain, and rinse out the equipment.

**Leader Guide**

# Colorful Lather Printing

> Suggestions for leading this activity begin on the next page.

In this activity, shaving cream (a soap lather) is used as a colorless base for supporting food color marbling patterns transferred to white paper. Campers marble paper with shaving cream and food color as they learn about the concepts of polarity and hydrophobicity. While this is a familiar activity to many educators, a new twist is added here—exploring how the colored shaving cream mixture behaves when a drop of water is added. You and your campers are sure to be surprised by the results. This new angle helps campers to further refine their mental pictures of the science of the system as well as the nature of soap, surfactants, solutions, colloids, and diffusion.

## Some Leader Background Information

Paper marbling is an ancient art, believed to have originated in the 12th century. In suminagashi (meaning "ink-floating"), a Japanese version of marbling, hydrophobic ("water hating") carbon-based inks are carefully dropped onto water and gently blown across the surface to produce swirls like those seen in polished marble. The ink is then picked up by applying rice paper to the surface of the water.

Another form of marbling, called ebru (meaning "colored wavy patterns on paper"), was developed in the 15th century in Turkey and Persia. Ebru uses a water-soluble gum as a thickening agent, making the water medium more viscous, while ox gall (a hydrophobic substance) is added to the surface to aid in the fluidity of the water-based inks that are applied. While these techniques involve patience and some relatively uncommon materials, you will discover in this activity that marbling can also be done with two common materials—shaving cream and food color.

During step 2 of this activity, the food color spreads into the paper due to wetting, surface tension lowering, and capillary action. After observing how food color spreads in water (step 3), campers discover in step 4 that color spreads in shaving cream less than in water because shaving cream is a lather. Its mixture of a liquid (soap dissolved in water) and tiny bubbles of the propellant gas (butane) might make it a foam, but the additional solid soap present makes it a lather.

The chemistry of soap is what makes the lather printing work. The soap in shaving cream has a water-loving (hydrophilic) "head" and a water-hating (hydrophobic) "tail." The water-based food color is attracted to the hydrophilic head of the soap and repelled by the hydrophobic tail. These two factors combine to limit the motion (or spreading) of the food color drops when added to the shaving cream in step 4. In this way the soap immobilizes the food color until the design is captured on the water-absorbent paper.

Once the shaving cream and food color are thoroughly mixed in step 8, the marbling action continues when a drop of water is placed on the surface of the tinted shaving cream and a white spot immediately forms. This spot results from the lowering of the surface tension in the dropped water at the point of contact. The food color originally present in the area is repelled (pushed out of the way) as the surface tension is lowered.

# LEADER

## Colorful Lather Printing
### Leader Guide

## Materials for Camper Procedure—See Left

## Procedure Notes and Tips

- Emphasize to campers that only one drop of food color is used in steps 2 and 3.
- Avoid piling too much shaving cream on the paper plate in step 4 to minimize the mess when shaving cream is scraped off the paper in step 6.
- Be sure campers put only drop-size amounts of food color on the shaving cream surfaces. Close monitoring of this step may be needed, depending on the abilities of the campers.

### Design your own...

As an extension to the activity, campers can design their own experiments to determine the effects of dropping other liquids in step 9. For example, try:

- cooking oil
- soda pop
- isopropyl alcohol

---

# CAMPER

## Colorful Lather Printing

### Overview

In this activity, you will use shaving cream, a common soap lather, to create beautiful colored patterns. At the same time you'll explore the chemistry of soap.

### Materials

✓ newspaper
✓ food colors in dropper bottles
✓ 3–4 pieces of a nonglossy, sturdy paper
  *Index cards, copy paper, or art paper work well.*
✓ small clear cup
✓ water
✓ aerosol shaving cream (standard white type)
✓ paper plate
✓ cooking spatula or craft stick
✓ toothpicks
✓ dropper or straw
✓ paper towels for cleanup

### Procedure

1. Cover your work surface with newspaper. (Food color can stain wood surfaces.)

2. Place one drop of food color onto a clean sheet of paper. Notice the degree to which the color spreads on the paper.

3. Fill a cup about half-full with room-temperature water. Without stirring, add one drop of food color to the water. Notice the degree to which the color spreads through the water.

4. Dispense a pile of shaving cream roughly the size of your fist onto the paper plate. Using a spatula or craft stick, shape the shaving cream so that the top surface is nearly flat and the area of this top surface is slightly larger than the paper that you will be marbling. Apply drops of several different colors of food color to different locations on the shaving cream. (A total of 6–8 drops works well.) Notice the degree to which the color spreads through the shaving cream.

# Colorful Lather Printing

## Leader Guide

**Q1:** In step 2, the color spreads into the paper. The spreading of the color in the shaving cream (step 4) is less than that observed in water (step 3).

**Q2:** Mousses, whipped cream, some hand soaps, and carpet cleaners are similar examples of colloids. Colloids contain larger particles than solutions. Solutions and colloids are both mixtures of two or more chemical substances.

## Wrap-Up

Ask campers to discuss their results and answers to the questions. Explain the hydrophilic and hydrophobic nature of shaving cream and why a white spot forms on the tinted shaving cream in step 9. (For help with the discussion, see Leader Background Information.)

---

### CAMPER

### Camper Notebook

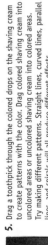

5. Drag a toothpick through the colored drops on the shaving cream to create patterns with the color. Drag colored shaving cream into uncolored areas or uncolored shaving cream into colored areas. Try making different patterns. Straight lines, curved lines, parallel lines, and spirals will all produce different effects.

6. Press the paper onto the surface of the shaving cream. You will notice that the paper becomes wetted by the colored shaving cream, and some of the color pattern may show through the paper. Pull the paper off the shaving cream. Some shaving cream will adhere to the paper. Scrape off the excess shaving cream close to the paper using the spatula (or craft stick) and return the excess shaving cream to the original pile. The patterns of color that you created on the surface of the shaving cream should transfer to the paper.

7. You can make additional marbled papers by repeating steps 4–6 if you wish. Otherwise, move on to step 8.

8. Using the spatula (or craft stick), mix the pile of colored shaving cream until it is one uniform color. If the color is very pale, mix in a few more drops of food color.

9. Apply a single drop of water to the surface of the colored shaving cream and observe what happens. You may wish to apply additional drops of water at different places on the surface for design purposes. Now try repeating steps 5–6 with the shaving cream mixture that remains.

 *Even though shaving cream is formulated to be safe on skin, it can become irritating if left on your skin for too long. Be sure to wash your hands when you are finished with this activity.*

**Q1:** Describe the spreading you observed when dropping food color onto clean paper (step 2), into water (step 3), and onto shaving cream (step 4).

**Q2:** Shaving cream is a lather, similar to a foam. A foam is a colloid (a gas trapped within a liquid). Do research to answer these questions: What other common products are foam or lather colloids? How are colloids, in general, different from solutions? What do solutions and colloids have in common?

# LEADER

## Surfactants

## Leader Guide

> Suggestions for leading this activity begin on the next page.

Campers learn how surfactants, such as soaps and detergents, affect the surface tension of water. This change in surface tension is important in how soap and detergent molecules entrap dirt and germs that collect on the skin and get flushed down the drain during the rinse.

## Some Leader Background Information

A surfactant is a chemical that increases water's ability to wet or spread out over a surface. Surfactants do this by reducing the water's surface tension. Surface tension is the property of water that causes drops of water to form sphere-shaped beads. Soaps and detergents are both surfactants.

Dirt and grime on your skin or clothes are usually contained in a layer of oil. It's this oil that makes it a challenge to wash things with water alone, because oil and water don't mix. The trick to cleaning is to use something that works with the water to pick up the oil. That's where soaps and detergents come in. They are molecules that have one end that is attracted to the oil and another end that is attracted to water. While you wash, the soap molecules' water-hating tails cluster around oil droplets to form

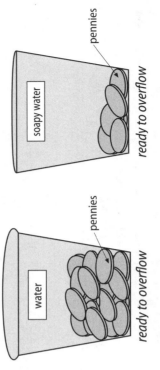

*soap or detergent molecule*

*micelle*

micelles. Dirt and oil are embedded (trapped) in these micelles and get flushed down the drain during the rinse because the water-loving heads are attracted to the water.

In this activity, campers discover that purified water has a very high surface tension. This results from the very strong attractions of water molecules to each other. Water molecules on the top of the first cup (holding just purified water) are more attracted to the water molecules inside the cup than they are to the surface (inside) of the cup or the air above the cup. This causes the water to bead up and form a dome when the counters are added.

Soaps and detergents added in the water lessen the strong attractions of water molecules toward each other. Since the water molecules on the top of the soapy water cup are attracted more strongly to the cup surface and air and less strongly to the inside water molecules, the water surface of the cup containing soapy water is flatter when the counters are added. Less counters are needed to cause the cup to overflow.

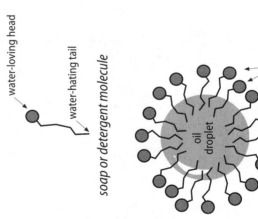

*ready to overflow*     *ready to overflow*

Water molecules do the same thing on top of the pennies. Less soapy water fits on the top of the penny because the water contains a surfactant.

# Surfactants

## Leader Guide

### Materials for Camper Procedure—See Left

### Supply Information

- Distilled water is available at grocery stores.

### Materials for Getting Ready

- ✓ clean 1- or 2-L bottles
- ✓ distilled water, deionized water, or purified drinking water without added minerals
- ✓ dishwashing liquid
- ✓ 5-mL pipet or teaspoon

### Getting Ready

1. Calculate how much soapy water is needed for campers to do the activity. One liter will make a little over 4 cups. To each liter of purified drinking water, add 5 mL (1 teaspoon) dishwashing liquid. Fill the bottles and label them as follows: soapy water containing _____ (5 mL/L) fill in detergent name

2. Experiment with the type of cup and water campers will be using to determine the approximate number of counters each camper (or group of campers) will need. For example, a 3-ounce bathroom cup filled with purified water may require 15–20 pennies before overflowing, a 9-ounce short SoloGrips™ plastic cup may require 35–50 pennies, and a 10-ounce tall plastic cup may require 20–30 pennies.

### Procedure Notes and Tips

- If time is short, you can have some campers do steps 1–4 and other campers do steps 5–7.

### Design your own...

As an alternative to the provided activity, campers can design their own experiments. For example:

- Determine the variables that affect the number of water drops that can be placed on a coin.

Page 151

---

# Surfactants

## Surfactants

### Overview

Why do soaps and detergents belong to a group of chemicals called surfactants? Experiment with soapy water to find out.

### Materials

- ✓ 2 new plastic cups (free from any soap or detergent residue) such as 3-ounce (90-mL) bathroom cups or 9-ounce (270-mL) tumblers
- ✓ same-sized counters that sink in water (such as pennies or plastic craft beads)
- ✓ distilled water, deionized water, or purified drinking water without added minerals
- ✓ soapy water (prepared by leader)
- ✓ dropper or disposable pipet
- ✓ clean surface (such as waxed paper or bottom of plastic cup)
- ✓ 2 pennies
- ✓ paper towels for spills and clean up

### Procedure

#### Test with Cups of Water

1. Fill one new, clean cup to the very top with purified water. In the data table at the end of this activity, sketch a side view of the cup to show what the top surface of the water looks like before adding the counters in step 2.

2. Gently add counters to the cup one at a time without splashing the water. In the data table, record the actual number of counters you added just before the water in the cup spills over the side. Sketch a side view of the cup to show the shape of the surface of the water after adding the counters and just before the water spills over the side.

   **Q1:** What happens to the shape of the surface of the water?

3. Fill another cup to the very top with soapy water. In the data table, record the name and concentration of the detergent solution. Sketch the top surface of the water.

Page 44

# LEADER

## Surfactants

### Leader Guide

- Campers can place their empty cups and pennies on paper towels before filling them with water. This step not only helps with cleanup, it also makes the overflow of liquid more apparent.

- If the cups are not completely filled with water before the counters are added, larger amounts of counters will be needed.

**Q1:** The water forms a dome on the surface.

**Q2:** The surface of the soapy water is flatter than the surface of the water with no detergent.

**Q3:** Answers will vary, but campers could experiment with different brands of detergents and different amounts of detergent to determine if the number of counters will change. They would discover that the amount of detergent needed for the critical micelle concentration is very small. (For example, results with soapy water containing 5 mL/L, 2 mL/L, and 1 mL/L detergent show that approximately the same number of drops of these soapy solutions fit on the head of a penny.)

**Q4:** The penny holding just water forms a taller, rounder water dome than the penny holding soapy water.

**Q5:** Answers will vary, but one observation might be that a raindrop on a car domes up rather than spreads flat.

### Wrap-Up

Have campers share their results. Define the terms "surfactant" and "surface tension." (For help with the discussion, see Leader Background Information.) Describe how surfactants work to remove dirt and germs from skin and clothing.

---

## CAMPER

### Camper Notebook

4. Repeat step 2 with the soapy water cup.

**Q2:** What happens to the shape of the surface of the soapy water?

**Q3:** Do you think the number of counters that can fit in the cup containing soapy water depends on the type and brand of detergent and the amount of detergent used to make the soapy water? How can you find out?

**Test with Pennies**

5. Fill a clean dropper or disposable pipet with purified water. Drop one drop on a clean surface such as a piece of waxed paper or the bottom of an upside down plastic cup. After observing the size of one drop, predict how many drops of purified water you think will fit on the heads side of a penny. Write your prediction in the data table at the end of this activity.

6. Counting as you drop, use the dropper to carefully place drops of purified water onto the heads side of a penny. In the data table under trial 1, record the actual number of drops added just before water spills off the penny. Dry the penny and repeat for two more trials. Calculate an average.

7. Place a second penny heads up. Fill a dropper with soapy water. In the data table, record the name and concentration of the detergent used and your prediction of how many drops of soapy water will fit on the penny. Then, counting as you drop, carefully place drops of soapy water on the heads side of the penny. In the data table under trial 1, record the actual number of drops added just before the water spills off the penny. Dry the penny and repeat for two more trials. Calculate an average.

**Q4:** Describe what is different about the shape of the liquid surfaces on the two pennies.

**Q5:** Have you observed the high surface tension of water in other instances before today? Where?

# Leader Guide

## Surfactants

### Camper Notebook

**Test with Cups of Water**

| | Sketch Top Surface Before Adding Counters | Sketch Top Surface Just Before Water Spills Over |
|---|---|---|
| cup of purified water | | number of counters _____ |
| cup of soapy water containing _____ detergent name [ _____ detergent concentration ] | | number of counters _____ |

**Test with Pennies**

| | Predicted Number of Drops | Actual Number of Drops | |
|---|---|---|---|
| penny holding purified water | | trial 1 | |
| | | trial 2 | |
| | | trial 3 | |
| | | average | |
| penny holding soapy water containing _____ detergent name [ _____ detergent concentration ] | | trial 1 | |
| | | trial 2 | |
| | | trial 3 | |
| | | average | |

Disease Control Is in Your Hands—Surfactants

# LEADER

## Blowing Bubbles

### Leader Guide

👉 **Suggestions for leading this activity begin on the next page.**

This activity serves as a fun science extension for the collection of activities on soaps and detergents. Campers determine which brand of dishwashing detergent forms the biggest soap bubbles. In the Wrap-Up, campers learn details about the structure of soap bubbles.

### Some Leader Background Information

In a soap bubble, two layers of soap (or detergent) sandwich a water layer. The soap molecules orient themselves so their water-loving heads are toward the inside water layer, while their water-hating tails are oriented toward the outside. These water-loving heads and water-hating tails are important in both the structure of a bubble film and the ability of the soap to remove oily and greasy dirt from skin, fabric, or other surfaces.

The thickness of a soap bubble's water layer can change over time due to gravity and evaporation. When light passes through this changing thickness, interference patterns create multiple-colored patterns. In this activity, glycerin (a thick, syrupy, water-loving substance) stabilizes the soap bubbles by delaying some evaporation of the water layer.

There are different ways that this activity can be presented to campers. One way is to have campers follow the procedure in the Camper Notebook. Another way to do this activity is to demonstrate some parameters for blowing bubbles and then ask campers to devise their own experiments for evaluating the detergents' bubble capacity. Before campers begin brainstorming about their experiments, you may want to introduce or review the characteristics of good experiments.

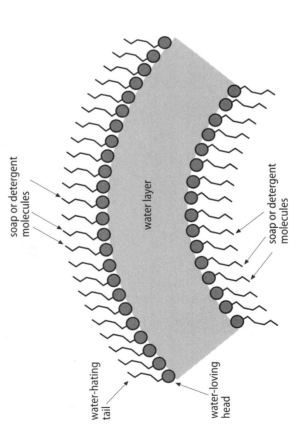

*Cross section of bubble film*

# Blowing Bubbles

## Leader Guide

### Materials for Camper Procedure—See Left

### Materials for Getting Ready

- ✓ 3 containers (with lids) capable of holding at least 500 mL (about 2.5 cups)
- ✓ 1500 mL (6 cups) water
- ✓ 500-mL graduated cylinder or liquid measuring cup
- ✓ 25 mL (5 teaspoons) each of 3 different dishwashing detergents
- ✓ 25-mL graduated cylinder or teaspoon
- ✓ 21 drops glycerin
- ✓ dropper
- ✓ plastic cups for dispensing bubble solutions to campers
- ✓ (optional) wading pool, hula hoop that fits inside pool, cotton string, large cement block, and bubble solution to cover bottom of pool

### Getting Ready

1. Prepare three different bubble solutions at least one day ahead so bubbles that form on the surface just after mixing will not interfere with bubbles blown in the activity. For each bubble solution, put 500 mL (2 cups) water, 25 mL (5 teaspoons) dishwashing detergent, and 7 drops glycerin into a container with a lid. Label each container with the brand of detergent you used.

2. Swirl the solutions to mix well. Allow each bubble solution to sit overnight. Shaking the solutions rather than swirling them results in lots of little bubbles that must pop before the solutions can be used.

3. Gently pour the solutions into appropriately labeled separate plastic cups for each camper or group.

4. If you are going to do the giant soap bubble demonstration described in Introducing the Activity, prepare enough bubble solution and set up the wading pool ahead of time. Wrap the hula hoop with cotton string.

---

## CAMPER

## Blowing Bubbles

### Overview

Soap bubbles make beautiful vessels and discover which dishwashing detergent forms the biggest soap bubbles.

### Materials

- ✓ 3 different kinds of dishwashing detergent solutions in plastic cups (prepared by leader)
- ✓ teaspoon or 5-mL graduated cylinder
- ✓ 3 large trays (such as cookie sheets lined with black plastic or large dark-colored Styrofoam® meat trays)
- ✓ 3 plastic coffee stirrer straws or drinking straws
- ✓ metric ruler or meterstick
- ✓ paper towels
- ✓ vinegar in spray bottle
- ✓ (optional) spot markers such as paper clips

*Blow a large bubble*

*Measure the wet ring*

### Procedure

1. Pour 1 teaspoon (5 mL) of the first bubble solution into the center of a clean tray. Tip the tray to allow the bubble solution to completely cover the surface of the tray.

2. Dip one end of a coffee stirrer straw or drinking straw into the cup of bubble solution. Place the end of the straw so that it just touches the center of the soapy surface of the tray.

3. Gently and continually blow into the clean end of the straw. Try to blow one large bubble dome. Continue blowing until the bubble pops. The bubble will leave a wet ring on the tray that will be visible for at least a few seconds.

4. Use a metric ruler or meterstick to measure across the widest part of the ring left by the bubble. This is the diameter of the bubble. In the data table at the end of this activity, record the diameter of the bubble for trial 1 and the brand of the dishwashing detergent you used.

**Tip…** If the ring disappears faster than you can do step 4, use spot markers such as paper clips to mark each side of the widest part of the wet ring.

*Disease Control Is in Your Hands—Blowing Bubbles*

# Blowing Bubbles

## Leader Guide

### Introducing the Activity

You may want to introduce the activity by making a giant soap bubble.

- Cover the bottom of a wading pool with bubble solution.
- Use a hula hoop wrapped in cotton string as a giant wand that you slowly raise up from the bottom of the pool.
- For added fun, have a camper stand inside the pool on a large cement block before you raise the wand.

You may also want to demonstrate steps 1–3 of the procedure before campers try it.

### Procedure Notes and Tips

- Coffee stirrer straws work better than drinking straws because their thinner design is less disruptive to the bubbles. Drinking straws keep campers' faces further from the bubbles and offer the option of purchasing individually wrapped straws.
- A bubble forms more easily when the coffee stirrer straw is first dipped in bubble solution, as instructed in step 2. This step means that the amount of bubble solution on the tray will change during the course of the activity; however, this change should not compromise the results.
- If time permits during step 3, have campers try to blow at least three practice bubbles before recording data to get bubble sizes that are more representative of the performance of the bubble solution and less dependent on other variables (such as bubble-blowing skill).

### Design your own...

As an alternative to the provided activity, campers can design their own experiments. For example, campers can investigate the bubble-blowing ability of:

- different concentrations of the same detergent
- identical concentrations of different detergents
- the same detergent at different temperatures

---

## Camper Notebook

5. Repeat steps 3 and 4 at least two more times with the same bubble solution. Calculate and record the average bubble diameter for the bubble solution you just tested.

**Q1:** Are the sizes of the bubbles blown approximately the same? Why or why not?

6. Repeat steps 1–5 with the remaining dishwashing detergents. Make sure to use a different tray and straw for each solution tested.

**Q2:** Why do you use a different tray and straw for each solution tested?

**Q3:** Although consumers often think bubbles (suds) and lather are important for cleaning ability, bubbles and lather are not necessary for cleaning. Give an example of a cleaning situation when suds are desirable.

### To clean up...

- Throw away all straws.
- Wipe up any excess bubble solution from the tabletop with dry paper towels.
- Spray the surface with vinegar to remove the soap film and wipe the table dry with fresh paper towels. (Do not use water because that would create more suds.)

|  | Bubble Diameter | | |
|---|---|---|---|
| | Brand of Dishwashing Detergent: | Brand of Dishwashing Detergent: | Brand of Dishwashing Detergent: |
| trial 1 | | | |
| trial 2 | | | |
| trial 3 | | | |
| average | | | |

# Blowing Bubbles

## Leader Guide

- If time is short, different group members may study different dishwashing detergents. The group can then pool their data together to find the best dishwashing detergent formula for blowing big bubbles.

**Q1:** Answers will vary, but bubble size usually depends on blowing consistency and the skill of the person blowing.

**Q2:** Different trays and straws are used so that the different dishwashing detergents don't get mixed together.

**Q3:** Suds are desirable when consumers benefit from seeing that the cleaning product is working, such as during hand, hair, and car washing. Suds are undesirable when the cleaning process is contained within a machine, such as a dishwasher or clothes washer. Too many suds in these situations could cause the machine to overflow.

## Wrap-Up

Ask campers to compare and summarize their results. Some brands and formulations (such as Dawn® and Joy®) consistently produce bigger bubbles. Discuss the structure of soap bubbles. (For help with the discussion, see Leader Background Information.) If you want, campers can consider detergent advertisements and costs in light of their results.

# LEADER

# Watching Granny Smith Rot

👆 Permission is granted for you to copy and distribute this take-home activity for your event. (The master is provided in the Camper Notebook.)

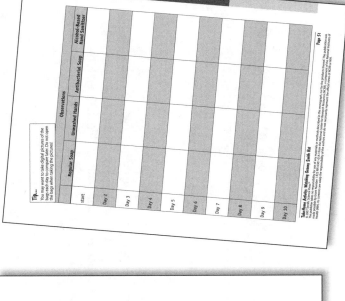

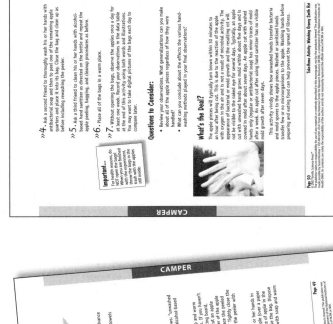

Disease Control Is in Your Hands—Watching Granny Smith Rot

## Leader Guide

# How Clean Is Your Clan?

> 👆 Permission is granted for you to copy and distribute this take-home activity for your event. (The master is provided in the Camper Notebook.)

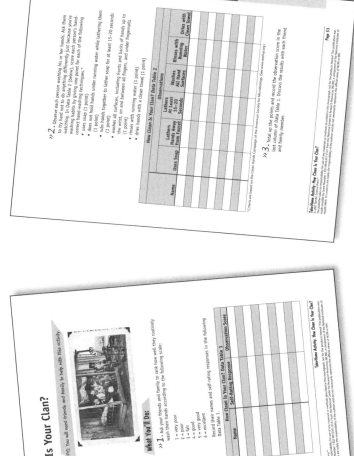
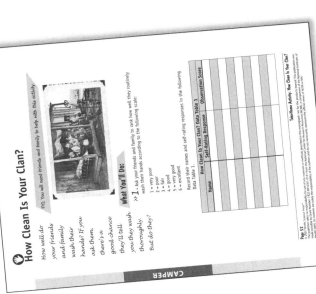

Disease Control Is in Your Hands—How Clean Is Your Clan?

LEADER

# TOPIC 2

## WATER PURIFICATION

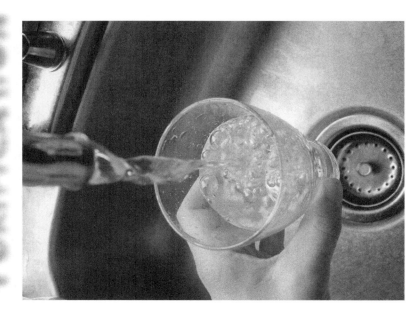

### List of Activities

✓ Back to Its Elements .................................................. 162
   *Observe how water can be broken apart into its component elements.*

✓ Distinguishing Water Samples ................................. 166
   *Investigate the differences between tap water and different bottled waters using super-absorbing crystals.*

✓ Osmosis with Eggs ..................................................... 170
   *Observe the process of osmosis using decalcified eggs. (Reverse osmosis is a common water treatment process.)*

✓ Water Taste Test ........................................................ 175
   *Do a water taste test with various brands of bottled water.*

✓ Investigating Mineral Content ................................. 179
   *Determine the mineral content of water samples based on their acidity or basicity.*

✓ How Hard Is Your Water? .......................................... 182
   *Estimate the relative amounts of dissolved minerals in different types of water.*

✓ Purifying Surface Water ............................................. 185
   *Investigate the role of flocculation in purifying water.*

✓ How Much Iodine Is Present? .................................... 189
   *Learn one method of monitoring chlorine levels in tap water. (Iodine is tested instead of chlorine because it is safer to handle.)*

✓ Removal of Iodine from Water .................................. 193
   *Explore the effectiveness of activated charcoal (commonly used in tap water purification) to remove residual chlorine. (Iodine is tested instead of chlorine because it is safer to handle.)*

✓ Take-Home Activity: What's Really in the Bottle? ... 197
   *Examine bottled water labels to learn the different types of bottled water.*

✓ Take-Home Activity: The Amazing Water Maze ..... 198
   *Take a journey through the water maze. What will you label the bottled water?*

**Page 161**

LEADER

# LEADER

## Back to Its Elements

**Leader Guide**

👉 **Suggestions for leading this activity begin on the next page.**

Campers observe how water can be broken apart into its component elements using an electric current.

### Some Leader Background Information

The chemical formula for water, $H_2O$, tells us that two atoms of hydrogen combine with one atom of oxygen to make one molecule of water. During electrolysis, water is broken down into its elements. The elements hydrogen and oxygen most commonly exist in their elemental forms as diatomic gases. A diatomic gas is a gas in which two atoms make up the gas molecule. In the cases of hydrogen and oxygen, two hydrogen atoms make up the hydrogen molecule and two oxygen atoms make up the oxygen molecule.

Hydrogen and oxygen are different molecules having different masses and sizes. In 1811, an Italian scientist named Amadeo Avogadro hypothesized that equal volumes of gases at the same temperature and pressure contain equal numbers of molecules. For example, equal volumes of hydrogen and oxygen gases contain the same number of molecules at the same pressure and temperature. This hypothesis, amply shown to be true over the last two centuries, is known as Avogadro's Principle. Thus, as water breaks down into hydrogen and oxygen gases, twice as much hydrogen is formed as compared to oxygen, and the hydrogen occupies twice as much volume.

Campers will notice that twice as much gas is collected in one electrode than in the other. The electrode attached to the positive battery terminal is the cathode. Hydrogen gas ($H_2$) forms at the cathode. The formation of $H_2$ at the cathode also produces hydroxide ions (OH−). The reaction at the cathode can be written as

$$4 H_2O + 4 e^- \rightarrow 2 H_2 + 4 OH^-$$

The electrode attached to the negative battery terminal is the anode, where oxygen gas ($O_2$) is formed. The formation of $O_2$ at the anode also produces hydrogen ions (H+). The reaction at the anode can be written as

$$2 H_2O \rightarrow O_2 + 4 H^+ + 4 e^-$$

Twice as much hydrogen gas is formed at the cathode than oxygen gas at the anode, since a water molecule contains two atoms of hydrogen and only one atom of oxygen. (The electrons that flow to or from the battery are not "seen," but are part of the reaction.)

The acid/base indicator in the electrolyte solution shows the presence of hydroxide and hydrogen ions. Hydroxide ions are basic and turn bromothymol blue indicator blue and purple cabbage juice indicator blue, green, or yellow. Hydrogen ions are acidic and turn bromothymol blue indicator yellow and purple cabbage juice indicator red, purple, or violet.

# Back to Its Elements

Leader Guide

## Materials for Camper Procedure—See Left

### Supply Information

- Floral wire is available in craft and hobby stores.
- Sodium sulfate can be obtained from chemical supply companies. As an alternative, Epsom salts (a soaking aid) are available in pharmacies.
- Solid bromothymol blue, phenolphthalein, or bromocresol green can be obtained from chemical supply companies.

### Materials for Getting Ready

✓ floral wire (iron wire painted green)
✓ wire cutters
✓ metric ruler
✓ balance capable of measuring to 0.001 g
✓ 1-L beaker
✓ electrolyte such as sodium sulfate ($Na_2SO_4$) or magnesium sulfate (Epsom salt, $MgSO_4 \cdot 7H_2O$)
✓ water
✓ acid/base indicator such as one of the following:
  - solid bromothymol blue, phenolphthalein, or bromocresol green
  - purple cabbage leaves, microwave-proof bowl, and microwave
  - purple cabbage leaves, blender, and strainer

### Getting Ready

1. Cut the floral wire into 10-inch (25-cm) lengths. Make two pieces for each apparatus.
2. If campers are using the straw electrodes but are too young to work with the hot-melt glue gun and glue, make the straw electrodes for them by following step 1 of the camper procedure.

Page 163

---

# Back to Its Elements

## Overview

Observe how water can be broken apart into its component elements using an electric current.

## Materials

✓ 2 small Beral pipets (disposable polyethylene transfer pipets) or transparent straw
✓ scissors
✓ metric ruler
✓ 2, 10-inch (25-cm) long floral wires (prepared by leader)
✓ hot-melt glue gun and glue (if using straws)
  *Use the hot-melt glue gun with adult supervision.*
✓ electrolyte and indicator solution (prepared by leader)
✓ petri dish or other small shallow bowl
✓ 9-volt battery
✓ goggles
✓ (optional) wire to make a stand
✓ (optional) tape
✓ (optional) alligator clips
✓ (optional) 9-volt snap connector and alligator clips

## Procedure

1. Prepare the electrodes in one of two ways.
   - Cut off the long stems of two pipets so that only about ¼ inch (0.5 cm) of the stems remain as shown at left. For each pipet, follow steps a–c.

   a. Push wire through pipet's closed end.
   b. Push until end of wire is even with pipet's cut end.
   c. Bend wire at 90° angle here.
   
   This end will be attached to a battery in step 3.

# LEADER

## Back to Its Elements
### Leader Guide

3. Prepare an electrolyte and indicator solution as follows:
   a. Mix 50–200 g sodium sulfate (or magnesium sulfate) into 1 L water.
   b. Add indicator to the solution by doing one of the following:
      - Add 0.001–0.005 g solid indicator to the 1 L (1000 mL) solution.
      - Make cabbage juice by tearing a couple of purple cabbage leaves and placing the pieces in a bowl with a very small amount of water. Microwave the mixture to boiling. Let cool to room temperature. Pour off and save the colored liquid. Add about 1–2 teaspoons (5–10 mL) or more of this liquid to the solution to darken it.
      - Chop purple cabbage leaves in a blender with a small amount of water and then strain the mixture to save the liquid. Add about 1–2 teaspoons (5–10 mL) or more of this liquid to the solution to darken it.

### Introducing the Activity

Explain to the campers that a molecule of water ($H_2O$) is made up of two atoms of hydrogen combined with one atom of oxygen. Also explain that electricity can be used to split water molecules into hydrogen and oxygen.

### Procedure Notes and Tips

- The reaction will start when the wires connect to the two terminals of the 9-volt battery. Make sure the wires do not make contact with each other.
- If campers are working in pairs, it is okay in step 3 to take turns holding the inverted electrodes upright in the container. If campers are working alone, it may be helpful for campers to construct a wire stand in step 3 to hold the inverted electrodes upright while connecting the wires to the battery terminals. The stand can be a wire

---

## CAMPER

### Camper Notebook

- As an alternative, cut a straw to make two 1½ inch (about 3–4 cm) lengths. For each straw, follow steps a–d.

  a. Push wire through the straw.
  b. Push until end of wire is even with end of straw.
  c. Seal straw at this end with hot glue. Let cool.
  d. Bend wire at 90° angle here.

  This end will be attached to a battery in step 3.

2. Fill each electrode with the electrolyte and indicator solution prepared by your leader. Cover the bottom of a container (petri dish or shallow bowl) with the same solution. A very shallow layer about ¼ inch (0.5 cm) deep is sufficient.

3. Quickly invert each filled electrode into the container's solution so the solution stays in the electrode and the electrode's long wire sticks out of the top as shown. Keep the electrodes upright at all times by either holding them, supporting them with a wire stand you make (Method A), or attaching them to the inside of the container with tape (Method B). Be sure the electrodes stay upright.

4. Connect one wire to each terminal of a 9-volt battery by one of the following methods:
   - Twist the wire around the terminal.
   - Twist the wire onto an alligator clip and clip the alligator clip to the terminal.
   - Twist the wire onto an alligator clip. Place a 9-volt snap connector on the battery. Clip the alligator clip to the wire coming off the connector.

   Most importantly, the electrode wires must not make contact with each other.

5. After the two wires are connected to the battery terminals, make observations for several minutes.

**Q1:** What changes are observed as the reaction proceeds?

**Q2:** How do the relative amounts of gases produced at each electrode indicate which gas is formed?

*Method A*

*Method B*

# Back to Its Elements

## Leader Guide

loop that snugly holds the two inverted electrodes and attaches to a loop surrounding the container. As an alternative, the electrodes can be taped to the inside wall of the container as long as the electrodes stay perfectly upright.

**Q1:** Bubbles rise from the ends of the wires and collect in the electrodes as the reaction proceeds.

**Q2:** The volume of gas collected in one pipet (straw) is twice the size of the other. Some campers may realize that, since water has twice as many hydrogen atoms as oxygen atoms, the larger volume of gas is hydrogen and the smaller volume of gas is oxygen.

## Wrap-Up

Discuss Avogadro's Principle. (For help with the discussion, see Leader Background Information.) Relate the results of the investigation to this principle by stating that we can deduce that the larger volume of gas is hydrogen and the smaller volume of gas is oxygen. Based on the campers' abilities, you may continue by describing the reactions that occur at each electrode. Talk about the production of hydroxide ions and hydrogen ions during the reaction. Explain that the acid/base indicator shows the presence of these ions.

# LEADER

## Distinguishing Water Samples

### Leader Guide

✋ *Suggestions for leading this activity begin on the next page.*

In this activity, campers use super-absorbing polymeric granules to investigate the amount of nutrients (dissolved ions and other substances) in various water samples. Campers then compare their results with the product labels to determine the accuracy (and the clarity) of the labels.

### Some Leader Background Information

When a water-containing liquid is added to the water-absorbent polymer crystals, the crystals swell as they absorb the water, forming gels with volumes that are much larger than the dry crystals. The amount of water the crystals absorb depends on the amount of time the crystals are in the liquid and the concentration of the ions that are dissolved in the liquid. In this activity, contact time between the crystals and the liquids is the same for each sample. However, the concentration of ions in the samples varies quite a bit.

The "purified" water samples (those labeled as distilled water or purified drinking water) have little or no dissolved ions in them; this accounts for the observation that the polymer crystals placed in these samples swell the most (even more than tap water). Spring and mineral water samples contain higher concentrations of ions than purified water; this accounts for the fact that the polymer crystals in these samples do not swell as much as those in purified waters. Gatorade and other sports drinks contain even larger concentrations of ions intended to replace those lost by the body during heavy exercise. The polymer crystals placed in these types of drinks absorb significantly less water than those in the other samples and swell only a very small amount.

The water-absorbent crystals used in this activity are made from sodium polyacrylamide, a polymer that contains ions. Sodium polyacrylamide

absorbs water molecules because water molecules are attracted to the ions in the polymer. Water molecules are also attracted to ions in the liquid samples tested in this activity.

The ions in the liquids and the ions in the polymer are in competition for the water molecules. The amount of water absorbed by the polymer depends on the concentrations of ions in the liquid. The more ions in the liquid, the less water molecules that can be absorbed by the polymer. Thus, the polymer swells less in liquids with high concentrations of ions.

The polymer crystals are somewhat similar to body cells, which bloat or absorb more water from a water solution high in water and low in ions. On the other extreme, body cells shrink or lose water in body solutions high in ions and low in water.

---

**How do ions get into water? When salts dissolve in water, they produce ions. Ions commonly found in tap water include sodium ion ($Na^+$), chloride ion ($Cl^-$), magnesium ion ($Mg^{2+}$), calcium ion ($Ca^{2+}$), fluoride ion ($F^-$), bicarbonate ion ($HCO_3^-$), and carbonate ion ($CO_3^{2-}$).**

# Distinguishing Water Samples

## Leader Guide

### Materials for Camper Procedure—See Left

### Supply Information

- Granular water-storing polymer products (such as Soil Moist™, Water-Gel Crystals, Sta-Moist™ Gel, Aquadiamonds®, or Watersorb®) are usually available where potting soil is sold.
- Distilled water is available in grocery stores.

### Procedure Notes and Tips

- Be sure campers analyze tap and bottled water as well as enhanced waters (such as sports drinks) having different amounts of minerals and other nutrients. This variation will make the trend more clear to the campers.
- Rather than have each camper test each water sample, you may want to divide the samples among the campers.
- If the time allocated to this investigation is less than 2 hours, use hot water samples in step 3 to hasten the absorption of water.

---

# Distinguishing Water Samples

## Distinguishing Water Samples

### Overview

Water-absorbent polymers in certain products (such as disposable diapers and water-storing crystals for potting soil) may absorb liquid many times their volume. How much liquid will these polymer crystals absorb? Is the amount of water absorbed by these crystals different for different kinds of bottled water and other beverages?

### Materials

- ✓ 9-ounce (about 270-mL) or larger clear cups (one for each liquid sample)
- ✓ permanent marker or self-stick labels
- ✓ granular water-storing polymer product having crystals measuring about 2–4 mm in diameter
- ✓ tap water
- ✓ 2 or 3 different types of bottled water (such as distilled water, spring water, and mineral water)
- ✓ Gatorade® or similar sports drink
- ✓ 1-cup (250-mL) liquid measuring cup with metric markings
- ✓ strainer (such as tea strainer or plastic cup with pushpin holes in the bottom)

### Procedure

1. Label a cup for each sample to be tested. Be sure to include tap water. In the data table at the end of this activity, record the names of the liquids you will test. Read the bottle labels and record the ingredient list for each sample.

2. Place 10 polymer crystals that are about 2–4 mm wide into each of the labeled cups. Since the crystals are irregularly shaped and some are closer to 2 mm while others are closer to 4 mm, try for a similar sample of 10 for each cup. (In other words, do not initially select only the largest crystals, leaving smaller crystals for the other cups.)

# LEADER

## Distinguishing Water Samples

### Leader Guide

**Q1:** Some campers will notice that the higher the nutrient content of the water sample, the lower the amount of water that the crystals absorb.

**Q2:** Answers will vary, but some campers may deduce that one or more of the bottle labels don't accurately or clearly indicate the amount of additives in the water. Also, some bottle labels may be very vague about their contents.

### Wrap-Up

Ask campers to share their results. Discuss how the super-absorbing polymer works. (For help with the discussion, see Leader Background Information.) Be sure campers understand that the more nutrients (dissolved ions) in the water sample, the less water absorbed by the polymer. Ask campers how the amount of water absorbed by the different water samples fits with what their labels say about the contents of the water.

---

# CAMPER

## Distinguishing Water Samples

### Camper Notebook

3. Add 150 mL of the appropriate liquid to each of the labeled cups. Allow the cups to sit several hours or overnight.

4. Hold the strainer over the measuring cup and pour the contents from one sample cup into the strainer. Once the liquid has drained into the measuring cup, return the crystals to their original (now empty) cup. (You are saving the crystals for later comparisons.) Read the volume of liquid you collected in the measuring cup in milliliters as accurately as possible. Record the volume in the data table. Rinse the liquid down the drain.

5. Calculate the volume of liquid absorbed by the crystals using the following equation:

   volume of liquid initially added to the cup − volume of liquid collected after straining = volume of liquid absorbed by the crystals

   Record your result in the data table.

6. Repeat steps 4 and 5 for each sample.

7. Examine the crystals that you returned to the cups in step 4. Are the crystals transparent (clear) or cloudy? Are air bubbles present? How do their swollen volumes compare? Record your observations in the data table.

8. In the data table, rank the typical swollen crystal size of each sample from smallest (number 1) to largest.

**Q1:** Look at the data that you collected. What (if any) trends do you observe with regard to the amount of liquid absorbed by the crystals and the ingredients/water sources listed for the samples?

**Q2:** Based on your experimental results and conclusions, which, if any, water samples have labels that are not accurate and/or clear? Describe the problems.

**Important...** When you are done with the crystals, throw them in the trash. Do not dump the crystals down the drain because they can clog plumbing.

# Distinguishing Water Samples

**Leader Guide**

## Camper Notebook

| Sample | Ingredients and/or Water Source Listed on Label | Volume of Liquid | | | Rank |
|---|---|---|---|---|---|
| | | Liquid Added to Cup (step 3) | Liquid Collected After Straining (step 4) | Liquid Absorbed by Crystals (step 5) | |
| tap water | Don't write anything. | 150 mL observations: | | | |
| | | 150 mL observations: | | | |
| | | 150 mL observations: | | | |
| | | 150 mL observations: | | | |
| | | 150 mL observations: | | | |
| | | 150 mL observations: | | | |
| | | 150 mL observations: | | | |

Water Purification–Distinguishing Water Samples

# LEADER

# Osmosis with Eggs

> Suggestions for leading this activity begin on the next page.

In this activity, campers examine the movement of water through the semipermeable membrane surrounding an egg. Osmosis and reverse osmosis require a semipermeable membrane—a membrane that lets some molecules (such as water molecules) pass through it but does not allow other substances (such as salts or sugar) pass through it. The inner "skin" of an eggshell is just such a membrane.

## Some Leader Background Information

A semipermeable membrane is basically a surface or film that allows the passage of some atoms or molecules through it, but not other atoms or molecules. Water can flow in either direction through a semipermeable membrane. If the semipermeable membrane is between two solutions of equal concentration, equal amounts of water will flow in each direction across the membrane and no observable change will result. In contrast, if the concentrations of the two solutions are different, water will flow through the membrane from an area of higher concentration of water to an area of lower water concentration. This process is called osmosis.

Many brands of bottled water are purified using a process called reverse osmosis. In reverse osmosis, pressure is applied to the side of the membrane where water is in lower concentration, forcing water through the membrane to the side of higher water concentration. In this way, water can be purified by pushing or forcing only water through the semipermeable membrane. The hard water ions, salts, and other impurities found in natural waters do not pass through the membrane.

## Leader Guide

In this activity, campers investigate semipermeable membranes and the process of osmosis by observing and measuring the movement of water in and out of an egg. The outermost layer of an eggshell is a thin layer of protein called the cuticle. Inside the cuticle is the main shell, which is composed of the mineral calcite (calcium carbonate) and protein. Inside the main shell are transparent membranes made up primarily of keratin, a protein also found in the external layers of skin, wool, hair, and feathers. The egg white contained within these membranes is composed of approximately 88% water, along with 11% protein and small amounts of other substances.

When an egg is soaked in vinegar (a solution of 5% acetic acid in water), the acid reacts with the hard calcite eggshell, slowly removing it. The egg with its shell removed is said to be decalcified. The transparent membranes remain on the egg. These membranes are semipermeable.

When a decalcified egg is placed in a corn syrup solution, water moves from the decalcified egg (through the membrane) into the corn syrup solution. The concentration of water inside the egg is greater than the concentration of water in the corn syrup. Because the net movement of water is out of the egg, the egg shrinks (losing both mass and volume). The sugar in the corn syrup solution will not move into the egg because the membranes are permeable to water molecules but not to sugar molecules.

The opposite observation is made when a decalcified egg is placed in water. In this case, the net movement of water is into the egg and the egg swells. The egg gains slightly in both mass and volume. Raw eggs placed in water will swell more than hard-boiled eggs because the movement of water into decalcified eggs is greater if the egg proteins have not been denatured by cooking.

# Osmosis with Eggs

**Leader Guide**

## Materials for Camper Procedure—See Left

## Materials for Getting Ready

✓ 3 raw eggs for demonstration
  *Due to breakage, you may want to prepare a few extra.*
✓ 2 hard-boiled eggs per group of campers
  *Due to breakage, you may want to prepare a few extra.*
✓ one or more containers (such as cups, beakers, and buckets) to hold all of the eggs
✓ vinegar (enough to cover all the eggs)
✓ (optional) heavy paper or light cardboard
✓ (optional) scissors

## Getting Ready

1. Put one raw egg aside for step 1 of the camper procedure. Soak the other two raw eggs and all of the hard-boiled eggs in vinegar until the shells are completely dissolved. Be sure the vinegar completely covers each egg. *This process will take several days.* Check the shells daily for evidence that the shells are close to being dissolved. To do this, rub the shells gently with a finger to see if the shells are loose enough to be rubbed off. If the shells are too thick to rub off without damaging the membranes inside the shells, discard the used vinegar and add fresh vinegar to submerge the eggs. When the shell appears to be almost completely dissolved (usually after two or three days), gently rub the eggs under running water or in a bucket of water to remove any remaining shell material. Be careful not to poke through the thin membranes.

2. Decide how campers will be measuring their eggs. If camp time is short and campers have to check their eggs the same day, the only method that will show appreciable results after 1½ hours is weighing the eggs (Method 1).

---

# CAMPER

## Osmosis with Eggs

### Overview

Most water must be treated before it is safe to drink. Reverse osmosis is one method for purifying water. What is osmosis and reverse osmosis? How do these processes work? In this activity, you'll use eggs to observe the effects of osmosis, and you'll research several other ways to purify water.

### Materials

✓ 2 large, uncracked hard-boiled eggs
✓ one or more of the following to measure the eggs:
  • balance capable of measuring to 0.1 g
  • egg template (provided at the end of this activity) copied on heavy paper or light cardboard
  • graduated cylinder at least 5.6 cm wide or cup large enough to totally submerge an egg in, a pan to set the cup into, and a teaspoon or other volume measure
✓ 2 cups or wide-mouthed containers large enough to totally submerge an egg in
✓ water
  *Tap water works okay, but bottled purified drinking water works better in step 3.*
✓ light corn syrup
✓ paper towels
✓ (optional) scissors

### Procedure

1. Observe closely as your leader puts an egg into vinegar.

   **Q1:** What evidence do you see that the shell is reacting with the vinegar?

2. Since the reaction of the eggshell and vinegar is slow, your leader has prepared the eggs you will use. Measure each egg by any or all of the following methods and record the results in the "Start" column in the data table at the end of this activity.

# LEADER

## Osmosis with Eggs

### Leader Guide

3. If campers will be using Method 2 to measure their eggs, copy the template master (located in the Camper Notebook) onto heavy paper or light cardboard. Make one template for each group of campers. To save activity time, you may want to cut out the circles for the campers.

## Procedure Notes and Tips

- For step 1 of the camper procedure, place a raw egg with an intact shell into a cup of vinegar. Allow campers to observe the reaction. Explain that the bubbling they observe is carbon dioxide gas that is produced as the vinegar reacts with the eggshell. Also show campers a decalcified raw egg to let them observe the thin and flexible semipermeable membrane. Explain to campers that they will use less delicate decalcified hard-boiled eggs. Discuss what you did to prepare the hard-boiled eggs for the campers.

- Before campers begin step 2, use one of the decalcified hard-boiled eggs to demonstrate how to measure the eggs using the same method or methods the campers will use. Remember that, if campers will be measuring their eggs after 1½ hours, they need to weigh their eggs (Method 1) to show appreciable results.

- If a more accurate volume is desired for Method 3 and a balance is available but volumetric glassware is not available, the mass of the displaced water can be measured and the volume calculated using the density of water.

- Perform the rest of the camper procedure along with the campers using the two decalcified raw eggs.

- If campers are unable to continue checking the eggs on multiple days, let the eggs soak in the cups (step 3) for 1½ hours before drying and measuring them using Method 1. The other measurement methods won't show enough change after 1½ hours, but they will show change after 24 hours.

# CAMPER

## Camper Notebook

**Method 1:** Weigh the egg carefully on a balance as accurately as possible. (This is the best method to use when you are going to remeasure after only 1½ hours.)

**Method 2:** Determine the approximate diameter across the short axis (width) of the egg. If the leader hasn't already done so, cut out the circles on the egg template. Select the smallest template hole that the egg will slip through on its own. Carefully put the egg through the hole using the same end of the egg each time (either the pointed end or the wide end). Have a hand underneath the template to catch the egg.

**Method 3:** Measure the volume of the egg by determining the amount of water it displaces. Do one of the following.

- Add water to a graduated cylinder about halfway up and record the volume. Add the egg and, making sure the egg is completely covered with water, record the second volume. The volume of the egg is the second volume reading minus the first volume reading.

- If you do not have a graduated cylinder, place a cup into a pan and fill the cup to the brim with water. Carefully add the egg to the cup, making sure that the egg is completely covered with water. The egg will displace a volume of water equal to its volume, and this water will overflow into the pan. Carefully remove the cup containing the egg without spilling any additional water into the pan. Use a teaspoon or other volume measure to determine the volume of water that spilled into the pan. This volume equals the volume of the egg.

3. Label one cup "water" and the other "light corn syrup." Place one egg in each cup. Completely cover the eggs with the designated liquids.

**Q2:** What do you predict will happen to the sizes of the two eggs over time? Write down your prediction.

# Osmosis with Eggs

## Leader Guide

- Explain osmosis after campers complete step 4. (For help with the discussion, see Leader Background Information.)

- As an option in step 4, after 72 hours rinse the egg that has been soaked in corn syrup. Place it in a cup of water so that it is fully submerged. Have campers predict what will happen to the size of the egg. Measure the egg after 24 more hours.

### Design your own...

As an extension to the activity, campers can design their own experiments to determine the effects of different liquids on the eggs. For example, try:

- Gatorade®
- carbonated soft drink
- isopropyl alcohol
- salt water designed for aquariums

**Q1:** Bubbles form as the vinegar reacts with the eggshell.

**Q2:** Answers may vary, but the egg in the corn syrup will get much smaller and the egg in the water will get a little larger.

**Q3:** Answers may vary. The egg soaked in water gains slightly in both mass and volume because the net movement of water is into the egg.

**Q4:** Water moves from the decalcified egg (through the membranes) into the corn syrup solution. The concentration of water inside the egg is greater than the concentration of water in the corn syrup. Because the net movement of water is out of the egg, the egg shrinks (losing both mass and volume).

---

## CAMPER

### Camper Notebook

4. After 24, 48, 72, and 96 hours, dry the eggs by gently patting them with paper towel. Measure the eggs using the same method you used in step 2 and record the results in the data table. If you can only check your eggs the same day, let the eggs soak in the cups (step 3) for 1½ hours before drying and measuring them using Method 1.

5. To help illustrate the results, graph the data for both eggs. (See the sample graph setup at left.)

6. After your leader has given you an explanation of osmosis, answer the following questions.

**Q3:** Which egg became larger in the process? Why do you think this happened?

**Q4:** In what direction does water move through the egg membranes when the egg is placed in the corn syrup? What's your evidence for this?

**Q5:** Research and list ways to purify water. Which methods involve a physical change and which involve a chemical change?

*Sample graph setup* (egg measurement vs. time)

### Egg Measurements

|  | Start | 1½ Hours (optional) | 24 Hours (1 day) | 48 Hours (2 days) | 72 Hours (3 days) | 96 Hours (4 days) |
|---|---|---|---|---|---|---|
| egg in water |  |  |  |  |  |  |
| egg in light corn syrup |  |  |  |  |  |  |

# LEADER

## Osmosis with Eggs

### Leader Guide

**Q5:** There are several ways to purify water, including:

- Distillation and condensation (boiling the water and condensing the steam) is a physical change.
- Filtration is a physical change.
- Chemical treatment (a chemical change) uses chemicals such as chlorine or iodine to kill microorganisms.
- Boiling (to kill microorganisms) is a physical change.
- Ion exchange is a chemical treatment that yields very pure water (the deionized water often used in chemistry laboratories). The ion-exchange reaction is a chemical change.
- Combining hydrogen and oxygen to make water is a chemical change.

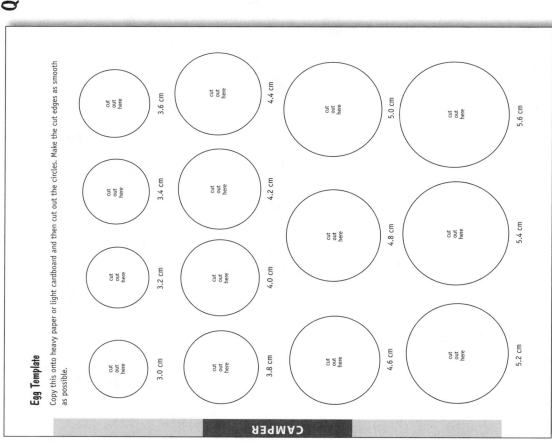

# Leader Guide

# Water Taste Test

👉 **Suggestions for leading this activity begin on the next page.**

This activity takes a close look at the variety of drinking water choices offered to consumers today. Many consumers are unaware of the sources and differences of drinking waters, even if they have a strong preference for a given water product. This taste test asks campers to rank the taste of different water samples without knowing their origins. When campers learn the identity of the samples, some may be surprised about which water sample they liked best.

## Some Leader Background Information

Tap water is usually purified water from either a surface water source (such as an ocean, lake, or river) or an underground water source (such as a spring or well). Bottled drinking water is usually additionally purified tap water. (The source may or may not be indicated on the bottle.) Natural spring water is from an underground source that naturally flows to the surface. (Whether the spring water is additionally purified depends on the purity of the water source.) Artesian water is water from an underground source under pressure. It may flow to the surface due to underground pressure or it may be pumped to the surface. Mineral water is spring or artesian water that naturally has at least 250 ppm dissolved solids. Sparkling water is water from a well or spring that naturally contains carbon dioxide; if treatment for impurities removes some of the carbonation, the carbonation lost can be replaced.

In this activity, it is unlikely that everyone will assign the same rank to all the waters. A majority of people may agree on which water is the worst tasting, but probably will not agree on which water tastes the best. Some people like water to have no taste at all, while others like a slight mineral taste, and so on. Think about what taste you prefer. Sparkling water is probably the easiest to identify based on its fizz. Campers may also be able to identify their local tap water because they are very familiar with its unique taste or odor.

In many blind taste tests, people prefer tap water to bottled water. In May 2001, the ABC Good Morning America show found that 45% of people tested preferred New York City tap water to Evian® or Poland Spring®. In England, one local water company found that 60% of 2,800 people surveyed couldn't tell the difference between popular bottled waters and the local tap water.

Campers may find little relation between the cost of the water and its ranking. Typically, bottled water is hundreds to thousands of times more expensive per ounce than plain tap water. Interestingly, some common soft drinks cost less per ounce than some popular bottled waters!

Water Purification—Water Taste Test

# LEADER

## Water Taste Test

**Leader Guide**

### Materials for Camper Procedure—See Left

### Materials for Getting Ready

✓ 5–6 different room-temperature water samples including tap water, mineral water, spring water, brand-name drinking water, and store-brand drinking water

✓ set of small, clear cups for each camper (one cup for each water tested)

☞ *Don't use paper cups because they transfer taste.*

✓ self-stick labels

✓ pencil or odor-free marking pen

☞ *Permanent markers should be avoided due to their strong odor.*

✓ paper plates or trays to transport the cups of water

✓ (optional) towels or paper towels

### Be sure to...

Keep track of the cost of each water sample. Many stores show unit prices (such as price per ounce) on the shelf near the product. If you need to calculate the price per unit volume,

- divide the price of the bottle in dollars by the number of ounces or milliliters in the bottle and
- multiply by 100 to get the price in cents per ounce or milliliter.

Assume that an average price for tap water is about 0.003 cents per ounce or 0.0001 cent per milliliter.

### Getting Ready

1. Label a set of cups for each camper using A, B, C, and so on.

2. Do not fill the cups in front of the campers. The identity of each water sample should be kept secret until step 6 of the procedure. After creating an answer key for yourself (you can use the table on the next page), pour a small amount of each water into the appropriate cup. Be sure that the same-lettered cup contains the same water sample for each camper (for example, cup A could contain everyone's spring water).

---

# CAMPER

## Water Taste Test

### Overview

Do you like tap water or a specific bottled water? In this activity, you will taste unidentified drinking water samples, rank them according to taste, and then attempt to identify them based on their appearance and taste alone.

### Materials

✓ cups of water samples (provided by leader)
✓ room where eating and drinking is permitted
✓ (optional) unsalted crackers to cleanse your palate between tastings

### Procedure

1. Look closely at each water sample. In the data table at the end of this activity, use a few words or phrases to describe the appearance of each sample.

2. Smell each water sample. In the data table, use a few words or phrases to describe the odor.

3. Sip each water sample. In the data table, use a few words or phrases to describe the taste. It is okay to retaste the samples. You may want to eat a piece of unsalted cracker between tasting samples to cleanse your palate.

4. Use your own opinions regarding taste to rank the water samples from best tasting to worst tasting, with 1 being the best. Record the ranks in the data table. Retaste each sample if necessary.

5. From the general list of water brands (or sources) provided by the leader, guess the identity of each water sample and enter your guesses in the data table.

6. From the leader, find out the actual water sample identities and prices per unit volume. Fill in the last two columns of the data table.

# Water Taste Test

## Leader Guide

| Water Sample | Brand or Source | Price per Bottle | Ounces or Milliliters in Bottle* | Price per Unit Volume* |
|---|---|---|---|---|
| A | | | | |
| B | | | | |
| C | | | | |
| D | | | | |
| E | | | | |
| F | | | | |

*Include units in table.

**3.** Prepare a general list that includes the brand (or source) of all the water samples you placed in the cups. Be sure to switch the order of the names around and do not identify which water sample goes with which cup. Campers will use this list in step 5.

**4.** Prepare a specific list that gives the letter of each cup and its corresponding water brand (or source). Include the price per ounce of each sample. Campers will use this list in step 6.

### Procedure Notes and Tips

- For safety purposes, this activity must be conducted outside of a laboratory since it involves drinking water samples. (Standard laboratory safety rules indicate "no eating or drinking in the lab.")
- All water samples should be at room temperature. Many people prefer cold water to room temperature water; however, chilling the water samples may influence the results.
- If any cup has to be refilled during the procedure, wrap a towel or paper towels around the bottle to hide its identity from the campers.
- You may want campers to repeat the entire activity with enhanced ("fitness") waters (such as Propel®, SmartWater™, and Fruit$_2$O®). For consistency in results, try to have campers test the same flavor of each water sample.

Page 177

## Camper Notebook

**Q1:** What, if any, results of the taste test are surprising?

**Q2:** What factors may influence the taste of the different waters?

| Water Sample | Appearance (step 1) | Odor (step 2) | Taste (step 3) | Rank (step 4) | Guess the Water Sample (step 5) | Water Sample Identity (step 6) | Price per Unit Volume (step 6) |
|---|---|---|---|---|---|---|---|
| A | | | | | | | |
| B | | | | | | | |
| C | | | | | | | |
| D | | | | | | | |
| E | | | | | | | |
| F | | | | | | | |

Page 66

## LEADER

# Water Taste Test

## Leader Guide

**Q1:** Answers will vary, but some campers may be surprised that they ranked tap water so high.

**Q2:** Answers will vary, but the taste of water may be influenced by chlorine in the water, minerals in the water, and how the water is stored.

## Wrap-Up

Have campers discuss their results with others. Show campers some bottled water labels and ask them to guess the source of each water based on the picture or graphic on the bottle's label. Now have campers read the bottle labels to find out where the water actually comes from. Point out that sometimes the images on the bottle have little resemblance to the actual water source. Also point out that some bottled waters don't reveal where their water comes from. Discuss the terminology used for water sources. (For help with the discussion, see Leader Background Information in this activity and see the fact sheet in the "Take-Home Activity: What's Really in the Bottle?")

**Leader Guide**

# Investigating Mineral Content

 *Suggestions for leading this activity begin on the next page.*

Campers may have spent some time considering the drinking water choices that today's consumers encounter. A person's preference might be based on cost, taste, or even the images from advertising. Can consumers tell the mineral content of bottled waters by reading the labels? In this activity, campers use paper washed with turmeric to analyze the mineral content of different drinking water samples and then assess the validity of bottled water labels.

## Some Leader Background Information

Most water samples are nearly neutral; this means that the water is neither very acidic nor very basic. A simple way to test the acidity or basicity of water is to use an indicator such as turmeric (a spice). Turmeric is an acid-base indicator that turns red in the presence of a base (such as baking soda), yellow in a neutral solution (such as distilled water), and yellow in the presence of an acid (such as vinegar). In this activity, campers observe what happens when various water samples are placed on turmeric-treated paper.

Water often contains magnesium carbonate and bicarbonate salts and/or calcium carbonate and bicarbonate salts (as well as other minerals). When a water drop is placed on the turmeric-treated paper, water slowly evaporates from the drop and also becomes absorbed by the paper so that the concentration of salts in the fibers of the paper increases. Carbonate and bicarbonate salts (such as baking soda) are bases so the paper turns red if these salts are present in the drop. The higher the concentration of these salts in the water drop, the darker red the spot turns on the paper.

In this activity, campers should discover the following:

- Distilled water contains the least amount of dissolved minerals.
- Waters purified by reverse osmosis should also have very few minerals, but some purified bottled waters (such as Dasani®) have minerals added after processing to improve taste.
- Spring waters may have more dissolved minerals than distilled water. However, not all spring waters will have the same amount of dissolved minerals because the amounts of calcium, magnesium, and other minerals in the water depend on the mineral composition of the aquifer from which the water came.
- The mineral content of tap water will vary from location to location, but typically falls in the same range as purified water and spring waters.
- Generally, mineral water has the most dissolved minerals.

# LEADER

## Investigating Mineral Content

Leader Guide

### Materials for Camper Procedure—See Left

### Supply Information

- Distilled water is available in grocery stores.
- Turmeric is sold as a spice in grocery stores.

### Materials for Getting Ready

✓ 2 plastic cups that hold at least 8 ounces (about 250 mL)
✓ water
✓ liquid measuring cup
✓ baking soda (NaHCO₃)
✓ teaspoon or other volume measure
✓ turmeric
✓ 99% isopropyl alcohol (rubbing alcohol)
☞ *Ninety-one percent isopropyl alcohol will also work, but its higher water content will increase the paper's drying time.*

### Getting Ready

1. Prepare a saturated baking soda solution by first adding ¼ cup (about 65 mL) tap water to a clean plastic cup. Add 1 teaspoon (5 mL) baking soda to the cup and swirl. The baking soda does not dissolve completely.

2. Prepare a turmeric solution by mixing 2 teaspoons (10 mL) turmeric with about ½ cup (130 mL) isopropyl alcohol in a clean plastic cup. The turmeric does not dissolve completely, but that's okay. If the alcohol evaporates from the cup over time, you can add more alcohol as needed.

---

## CAMPER

### Investigating Mineral Content

#### Overview

In this activity, you determine the relative mineral content of different water samples based on their acidity or basicity. Most water samples are nearly neutral; this means that the water is neither very acidic nor very basic. If a water sample is basic, this indicates the presence of minerals. A simple way to test the acidity or basicity of water is to use turmeric (a spice), which contains an indicator that is yellow if the sample is neutral or acidic and red if the sample is basic. In this activity, you will use turmeric-treated paper to test different water samples.

#### Materials

✓ manila file folder or other paper that is *not* acid-free
✓ turmeric solution (prepared by leader)
✓ paint brush, foam brush, or sponge
✓ distilled water in a container
✓ tap water in a container
✓ several different bottled waters
✓ saturated baking soda (NaHCO₃) solution (prepared by leader)
✓ vinegar (5% acetic acid)
✓ transfer tools such as droppers, disposable pipets, drinking straws, or cotton swabs (one for each water sample to be tested)
✓ (optional) hair dryer

#### Procedure

1. Dip a brush or sponge in turmeric solution and paint a long stripe on the left side of your paper. (See example at left.) Allow the solution to dry.

2. Write the name of each sample to be tested directly on the paper to the right of the turmeric stripe, separating the names along the stripe as much as you can. (See example at left.) Be sure to include the saturated baking soda solution and the vinegar as samples. Leave room to write additional information near each sample name.

*Example of setup*

distilled
tap—Grandma's well water
Aberfoyle—spring water
Evian—spring water
S. Pellegrino—mineral water
baking soda solution
vinegar

turmeric stripe

# Investigating Mineral Content

## Leader Guide

### Camper Notebook

3. For each bottled water sample, read the label or search the Internet to try to determine the source of the water (such as artesian, spring, mineral, sparkling, municipal, well, or purified) or its mineral content. (Some labels and Internet sites won't reveal this information.) Write down what you are able to find out next to the bottle's brand name. Also, write down where the tap water came from.

4. Using a dropper or other transfer tool, place 1 drop of each water sample on the turmeric stripe next to its name. To avoid cross-contamination, be sure to use a different transfer tool for each water sample.

5. Allow the water drops to dry. Air drying usually takes about 20 minutes. You can speed up the process by drying the drops with a hair dryer.

6. Record the color of the dried water drop next to the name of each water sample. Try to rank the samples from highest to lowest mineral content.

**Q1:** How many of the water samples you tested have high mineral content? Which ones are they? How did you determine this?

**Q2:** Look for patterns between the color of the dried drops and the water sources. Do any of the results appear to contradict what the water label says about either the source or mineral content?

**Q3:** How do the tap water results compare to spring and mineral water results?

**Remember...**
- The sample is neutral or acidic if its dried drop is yellow.
- The sample is basic if its dried drop is red.
- The darker red the dried drop, the more minerals the sample contains.

### Introducing the Activity

Explain how turmeric-treated paper works. (For help with the discussion, see Leader Background Information.) Be sure campers understand that minerals dissolved in water are bases. The higher the concentration of these minerals, the darker red the turmeric-treated paper will turn. Tell campers that the baking soda solution is basic and the vinegar is acidic.

### Procedure Notes and Tips

- Be sure campers test at least one mineral water.
- Make sure campers use a different transfer tool (pipet, dropper, drinking straw, or swab) for each water sample. You may want to keep one tool with each water sample so it can be used by all campers.
- You may want to have campers take home their turmeric-treated paper so they can test their home tap water.

**Q1:** Answers will vary depending on the water samples that are tested. Generally, mineral water has the most dissolved minerals. Campers can tell which of their water samples has higher mineral content because the spots left by these waters are darker red.

**Q2:** Results should agree with the statements on the bottle labels, but they may not due to false advertising. Results should, however, be consistent among all the campers. (See Leader Background Information for general trends.)

**Q3:** Results will vary depending on the tap water that is tested, but the results should be consistent among all the campers.

### Wrap-Up

Have campers share their results with the group. Have them discuss what they learned about each water sample.

# LEADER

## How Hard Is Your Water?

> Suggestions for leading this activity begin on the next page.

In this activity, campers use water hardness test strips to determine the relative amounts of calcium and magnesium ions in different types of water. These test strips are calibrated to measure water hardness. Distilled water (which has no dissolved minerals) is used as a reference.

## Some Leader Background Information

As rainwater passes through soil and underground rocks, it dissolves some minerals (such as calcium and magnesium ions) and other substances. Many bottled water companies fill their bottles with this groundwater by taking the water from wells or springs. Minerals usually remain dissolved when the water is bottled. Water that contains a large amount of dissolved minerals is called "hard" water. Some people prefer to drink water that has a higher mineral content.

Distilled water should have no dissolved minerals and a hardness of nearly zero. Waters purified by reverse osmosis should also have very low mineral content. However, some purified bottled water brands (such as Dasani®) have minerals added after processing to improve taste. Spring water may have more dissolved minerals than purified water. However, not all spring water will have the same amount of dissolved minerals because the amounts of calcium, magnesium, and other minerals in the water depend on the mineral composition of the aquifer from which the water came. Typically, mineral water samples have the most dissolved minerals. The mineral content of tap water will vary from location to location, but will typically fall in the same range as purified water or spring water.

Hard water poses no known danger to your health. In fact, hard water can be a good source of calcium and magnesium in the diet. However, hard tap water can be a nuisance because soap does not lather well and scaly

## Leader Guide

deposits are left behind on sinks, tubs, and glasses. Washing with hard water can cause dry, irritated skin and dry, tangled hair. Hard water can also cause problems in pipes, hot water heaters, and appliances such as dishwashers.

Soft water has lesser amounts of dissolved minerals and none of these problems. Water softeners convert hard water to soft by replacing the minerals that cause hardness. Usually a water softener exchanges calcium and magnesium ions (the hard water ions) with sodium or potassium ions.

# How Hard Is Your Water?

**Leader Guide**

## Materials for Camper Procedure—See Left

## Supply Information

- The water hardness test is often combined with other tests on a single strip. The test strips are available from aquarium, pool, and spa supply stores.
- Distilled water is available in grocery stores.

## Introducing the Activity

Discuss how groundwater becomes hard. (For help with the discussion, see Leader Background Information.) Encourage campers to read the bottle labels to try to determine the sources of the water when making their predictions in step 1. Point out that some bottles won't list their sources.

## Procedure Notes and Tips

- You can have campers work in groups so that they share the color chart included in the test strip package. (The color chart on the side of the package is needed in steps 3 and 4.)

---

**CAMPER**

# How Hard Is Your Water?

### Overview

As rainwater passes through soil and underground rocks, it dissolves some minerals (such as calcium and magnesium ions) and other substances. Many bottled water companies fill their bottles with this groundwater by taking the water from wells or springs. Minerals usually remain dissolved when the water is bottled. Water that contains a large amount of dissolved minerals is called "hard" water. Some people prefer to drink water that has a higher mineral content.

This activity allows you to determine the relative amounts of calcium and magnesium ions in different types of bottled water using water hardness test strips. These strips are calibrated to measure water hardness. Aquarium, swimming pool, and spa owners test their water with these strips to determine when water conditioning is needed.

### Materials

- ✓ small disposable cup for each water sample you are testing
- ✓ water hardness test strip for each water sample you are testing
- ✓ distilled water
- ✓ tap water
- ✓ several different brands of bottled water
- ⓘ *Be sure to include at least one spring water and one mineral water.*

### Procedure

1. Based on what you have learned about water sources and dissolved minerals, predict whether each water sample is soft or hard. List the water samples and write your hardness prediction in the data table at the end of this activity.

2. To test if your prediction is correct, pour a small amount of the first water sample into a clean disposable cup.

# LEADER

## How Hard Is Your Water?                                    Leader Guide

**Q1:** Answers will vary based on the waters tested. Typically, mineral waters are the hardest and distilled water is the least hard.

**Q2:** Answers will vary based on the tap water tested. The hardness of tap water depends on where it comes from, but it typically falls in the same range as purified or spring waters.

**Q3:** Answers will vary based on the spring waters tested. The amount of dissolved minerals in spring water depends on the mineral composition of the aquifer from which the water came. Therefore, not all spring waters have the same hardness.

## Wrap-Up

Ask campers to share their results and conclusions about the relationships between water source and hardness.

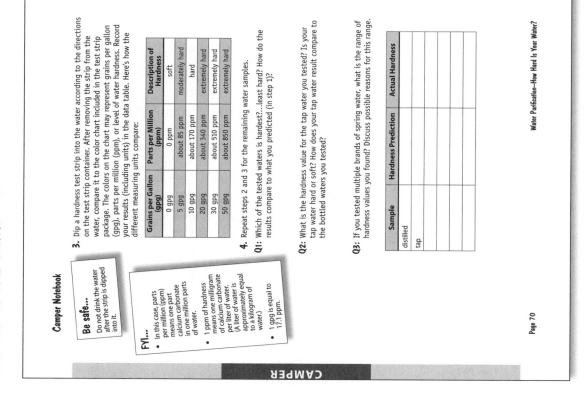

Water Purification—How Hard Is Your Water?

## Leader Guide

# Purifying Surface Water

> ✋ Suggestions for leading this activity begin on the next page.

One big advantage of drinking tap water rather than bottled water is that tap water is much less expensive. However, municipal tap water could be from surface water such as a river or lake. Campers may say they like the idea of drinking water from a source pictured or implied on a bottled water label rather than from a known river or lake that is slightly brown and may be polluted. This activity looks closely at flocculation and its role in purifying water. It also helps campers understand the processes behind producing safe tap water.

## Some Leader Background Information

Alum (aluminum potassium sulfate) is a water soluble aluminum salt. When alum dissolves in water, the resulting solution is slightly acidic. If the solution is made neutral by the addition of a base (like baking soda or ammonia), insoluble aluminum hydroxide forms. This aluminum hydroxide precipitate is somewhat unique because it is very fine, but the small particles that are formed attract each other and stick together. Since the small particles stick together, the larger particles formed can more easily be separated by filtration. The process of forming this precipitate is called flocculation.

When flocculation is used to purify surface water containing finely dispersed soil and clay, aluminum hydroxide particles stick to these small impurities to form larger particles that can be filtered out more easily. Aluminum hydroxide is also capable of trapping some bacteria, allowing some microbes to be filtered out. Water treatment facilities often use beds of fine sand to filter water.

In this activity, two different bases may be used to form the aluminum hydroxide. If baking soda is used, one drawback is that the aluminum hydroxide is produced all at once in a very fine, voluminous white precipitate. Household ammonia can also be used as a base in place of baking soda. The aluminum hydroxide forms more slowly when ammonia is used as the base. Campers can see the aluminum hydroxide stick to itself as more of the flocked precipitate forms and comes out of solution. One drawback to using ammonia as the base is that campers need to wear safety goggles because ammonia can damage eye tissue.

Whether baking soda or ammonia is used as the base, the formation of insoluble aluminum hydroxide is observed during step 5 in the tap water as well as in the surface water sample. The effect of the addition of aluminum hydroxide to the surface water sample is most dramatic after filtration in step 6. While the coffee filters do not trap all of the fine clays and precipitates, the filtered surface water treated with aluminum hydroxide is noticeably less yellow or brown than the filtered surface water sample without aluminum hydroxide treatment.

# LEADER

## Purifying Surface Water

**Leader Guide**

### Materials for Camper Procedure—See Left

### Supply Information

- Try to use granulated alum instead of powdered. Alum is sold in the spice section of grocery stores.

### Introducing the Activity

Tell campers that they will be making a dirty water sample that simulates surface water that may come from rivers and lakes. Campers will then treat this water with alum and baking soda (or ammonia) to mimic aluminum hydroxide treatment used by some municipalities to purify surface water. Be sure to tell campers not to drink any water samples in this activity.

### Procedure Notes and Tips

- As an alternative to step 1, obtain water from a local river, lake, or pond.
- Be sure campers stir the alum in step 4 until it is completely dissolved. It may not be possible to observe when the alum has dissolved in the dirty surface water (due to the finely dispersed dirt and clay). Therefore, campers first observe how long alum takes to dissolve in tap water and then they stir the surface water mixture just as long.
- Baking soda can be used in step 5 without safety goggles, since baking soda is used in the kitchen. If household ammonia is used instead of baking soda, make sure campers use safety goggles since ammonia is potentially damaging to eye tissue.
- Since the materials for flocculation are cheap, you may want campers to see how "clean" the surface water sample will become after a second or even third aluminum hydroxide treatment followed by filtration. If so, have campers repeat steps 4–6 one or two more times.
- Even if some of the water samples look clean, caution campers that they must not taste any water samples in this activity.

Page 186

---

# CAMPER

## Purifying Surface Water

### Overview

If your tap water comes from surface water such as a river or lake, you may wonder how the water is purified. Surface water sometimes looks dirty, cloudy, and yellow to brown in color. This activity shows you one way that surface water is treated to improve its quality.

### Materials

- soil
- paper to make a funnel
- 1 tablespoon and ⅛ teaspoon or other volume measures
- empty water bottle with a cap
- self-stick labels or permanent marker
- tap water
- liquid measuring cup
- 8 transparent plastic cups that hold at least 9–10 ounces (about 270 mL)
- alum (aluminum potassium sulfate dodecahydrate, $KAl(SO_4)_2 \cdot 12H_2O$)
- plastic spoons for stirring
- baking soda (sodium bicarbonate, $NaHCO_3$)
- funnel (such as a funnel cut from a 2-L plastic soft-drink bottle)
- 3 coffee filters
- (optional) household ammonia ($NH_3$ in water) and safety goggles

⚠ Be sure to wear safety goggles when using ammonia because it can damage eye tissue.

### Procedure

⚠ Do not drink any water samples in this activity. Samples that look clean may contain harmful bacteria that could make you sick.

1. Use a paper funnel to place 2 tablespoons (30 mL) of soil in an empty water bottle. Label the bottle "surface water." Add tap water to fill the bottle, put on the cap, and shake to mix. This mixture represents dirty surface water. It is okay if stones and heavy soil particles sink to the bottom of the bottle.

Page 71

# Purifying Surface Water

## Leader Guide

**Q1:** A solid forms in both cups when the baking soda (ammonia) is added.

**Q2:** The sticky aluminum hydroxide solid causes fine solids to clump together in the surface water sample. When these solids are filtered out, the water looks cleaner. When the untreated surface water sample is filtered, some solid is removed but the water still looks dirty. Adding aluminum hydroxide and filtering the water works better than just filtering the water.

## Wrap-Up

Ask campers to discuss their results. Tell campers that this activity demonstrates a water treatment process called flocculation. Discuss how flocculation works in removing some impurities and microbes from water. (For help with the discussion, see Leader Background Information.) Tell campers that water treatment facilities often use beds of fine sand to filter water rather than use paper filters like the coffee filters used in this activity.

If you'd like, you or your campers can research a product called "PUR Purifier of Water." PUR uses the process of flocculation (along with coagulation and disinfection) to offer an inexpensive water treatment option for disaster areas and developing countries. One packet of PUR powder purifies up to 10 liters of contaminated water. After mixing contaminated water with the powder, filtering the treated water through a cotton cloth, and waiting 20 minutes to allow for complete disinfection, the water is safe to drink. According to the World Health Organization, this type of low-cost water purification can dramatically improve the quality of stored household water and greatly reduce diarrheal disease, the leading killer of young children in developing countries. PUR was developed through collaboration between the Procter & Gamble Health Sciences Institute and the U.S. Centers for Disease Control and Prevention.

---

# Purifying Surface Water

### Camper Notebook

2. Label cups as shown in the data table at the end of this activity.
3. Fill the cups as listed in the data table.
4. Add alum as listed in the data table. Stir until all of the alum dissolves. If particles in the surface water prevent you from observing when all of the alum has dissolved, stir the mixture as long as you stirred the tap water mixture.
5. Add baking soda as listed in the data table. As an alternative, you can put on safety goggles and add ⅛ teaspoon (½ mL) household ammonia instead of the baking soda. Be sure to wait 3–5 minutes before continuing with step 6.

**Q1:** Describe what happens in both cups when you add the baking soda.

6. Label cups and filter the mixtures as listed in the data table. Each time you filter, place a clean coffee filter in the soft-drink bottle funnel and place the funnel on top of a new cup. Allow the mixture to drain into the cup.
7. Compare the physical characteristics of each water sample. Record your observations in the data table.

**Q2:** What is the result of adding aluminum hydroxide (alum and baking soda) and filtering the dirty surface water?

- coffee filter
- soft-drink bottle funnel
- cup

# LEADER

## Purifying Surface Water

### Leader Guide

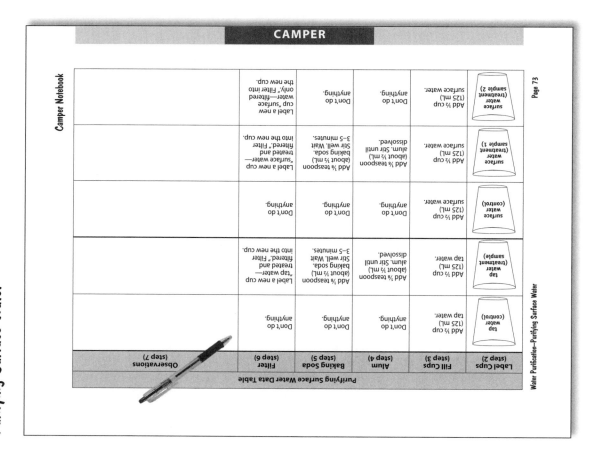

**Leader Guide**

involves multiplying the average number of drops of vitamin C it took to decolor the water sample by $1\times10^{18}$ (1 billion billion or 1 billion$^2$). This $1\times10^{18}$ adjustment factor is calculated as follows:

- The vitamin C solution concentration (500 mg per 100 mL) is $1.7\times10^{22}$ particles/L. In particles of vitamin C per drop, this is $0.85\times10^{18}$ particles vitamin C per drop or approximately $1\times10^{18}$ particles of vitamin C per drop. This value is obtained from 1000 mL per L and 0.05 mL per drop (20 drops per mL) and is shown in the following equation:

$$\frac{1.7 \times 10^{22} \text{ particles}}{1\,L} \times \frac{1\,L}{1000\,\text{mL}} \times \frac{0.05\,\text{mL}}{\text{drop}} = \frac{0.85 \times 10^{18} \text{ particles}}{\text{drop}}$$

- Since one particle of vitamin C reacts with one particle of iodine ($I_2$), $1\times10^{18}$ particles of iodine are present for each drop of vitamin C solution that is added to decolorize the water (iodine-containing) sample. Remind campers that the adjustment factor to multiply the number of drops is a very, very large number since the particles of iodine and the particles of vitamin C are both very, very small.

Some additional comments:

- If campers are unfamiliar with scientific notation, remember that $10^{18}$ is $10^9 \times 10^9$, a billion$^2$, or a billion billion. Campers can also use the billion billion or the exponent billion$^2$ as a unit or label.

- The particles of vitamin C per L (used in the calculation above) is calculated as follows:

(1000 mg = 1.000 g and 100 mL = 0.100 L)

$$\frac{0.500\,\text{g vit. C}}{0.100\,L} \times \frac{1\,\text{mole vit. C}}{176.12\,\text{g vit. C}} \times \frac{6.02 \times 10^{23}\,\text{particles vit. C}}{1\,\text{mole vit. C}} = \frac{1.7 \times 10^{22}\,\text{particles}}{1\,L}$$

- If the approximation of "$0.85\times10^{18}$ particles of vitamin C per drop is about $1\times10^{18}$ particles of vitamin C per drop" is not acceptable in your view, 0.5851 g vitamin C can be dissolved in 100 mL water to prepare a $1.00\times10^{18}$ particles of vitamin C per drop solution. (Alternatively, the 0.500 g vitamin C per 1000 mL water solution can be used with a $0.85\times10^{18}$ particles vitamin C per drop adjustment factor.)

# How Much Iodine Is Present?

> Suggestions for leading this activity begin on the next page.

Municipal water supplies are frequently sterilized with chlorine. Chlorine is highly toxic, so the right concentration is very important. Enough chlorine must be added to kill the microbes but the final chlorine concentration must not be harmful to people who drink and cook with the water. A closely related element that also kills microbes but is easier for campers to investigate is iodine. Iodine is also more highly colored than chlorine. In this activity, vitamin C is used to determine how much iodine is present in water samples. This process also demonstrates one way to determine chemical concentrations in water.

## Some Leader Background Information

Iodine solutions react with vitamin C. During the reaction, vitamin C (ascorbic acid) loses two electrons while being oxidized to dehydroascorbic acid. Iodine is reduced to the colorless iodide ion by gaining two electrons.

$$I_2 + \text{ascorbic acid} \rightarrow 2I^- + \text{dehydroascorbic acid}$$

Since one iodine molecule reacts with one vitamin C molecule, the amount of vitamin C that must be added to completely decolorize an iodine solution determines the amount of iodine present.

A microscale (small-scale) titration can be used to demonstrate this reaction. Campers measure 20.0 mL amounts (aliquots) of a light amber-brown iodine solution (their unknown water sample) and add a standard vitamin C solution dropwise until the iodine is reduced to colorless iodide ion. Campers should perform the titration at least three times and average the results to get a reasonably accurate estimate of the amount of vitamin C solution required to decolorize the water sample.

After the titration, campers need to calculate the approximate number of iodine particles present in their 20.0 mL water sample. This calculation

# LEADER

## How Much Iodine Is Present?

### Leader Guide

### Materials for Camper Procedure—See Left

### Materials for Getting Ready

✓ 10% povidone-iodine solution (such as Betadine®)

⚠ *The 10% povidone-iodine solution is available in drugstores as a topical antiseptic microbicide. Iodine itself is toxic and can cause irritation, stains, and burns to skin. However, iodine in the 10% povidone-iodine solution is nonirritating and nonstaining to skin and natural fibers. Be sure campers do not consume any iodine-treated solutions, either before or after activated charcoal treatment.*

✓ graduated cylinders capable of accurately measuring from 5 mL to 100 mL

✓ 3 or 4, 2-L graduated beakers or graduated cylinders

✓ 0.500 g (500 mg) vitamin C

✋ *If an analytical balance is available, use 0.5851 g vitamin C.*

✓ balance capable of measuring 0.001 g

✓ container that holds at least 100 mL

### Getting Ready

1. The solutions used in this activity are the same solutions that are used in "Removal of Iodine from Water." Prepare the following water samples. Label the samples A through C or D. The 2-L amounts will be enough for 75–100 campers to test one sample and share results (as outlined in the camper procedure) or for 25–30 campers to test each sample.

   - 5 mL povidone-iodine (PI) diluted to 2 L (2000 mL)
   - 10 mL PI diluted to 2 L
   - 20 mL PI diluted to 2 L
   - (optional) 50 mL PI diluted to 2 L

   ✋ *This sample requires a larger number of drops (about 20) during the titration.*

Page 190

---

# CAMPER

## How Much Iodine Is Present?

### How Much Iodine Is Present?

#### Overview

To kill harmful bacteria in drinking water, municipal water companies often add chlorine. Chlorine is very effective at killing bacteria; however, it is also toxic at high levels to plants and animals. How do water companies keep track of the amount of chlorine they add to water so people don't get sick? This activity illustrates one method for monitoring chlorine levels. However, you'll be using iodine (another microbe-killing element), because it is easier to handle.

#### Materials

✓ tap water
✓ graduated cylinder capable of accurately measuring 20.0 mL
✓ 3, 200-mL beakers or 3 clear plastic cups (at least 10-ounce size)
✓ white paper for a background
✓ water sample prepared by leader
⚠ *Do not drink the water sample because it contains iodine from a topical antiseptic.*
✓ disposable pipet, dropper, or small plastic squeeze bottle with dropper dispenser
✓ vitamin C solution prepared by leader

#### Procedure

1. Measure 20.0 mL tap water and pour it into a beaker or cup. This is the clear control. Place the container on a white sheet of paper to use for the color comparison in step 3.

2. Measure 20.0 mL of your assigned water sample and pour it into another beaker or cup.

3. For trial 1, add one drop of the vitamin C solution to your water sample. Swirl to mix well. Compare your water sample to the clear control. If the water sample is not completely decolorized after mixing, add another drop of the vitamin C solution and mix well. Count the number of drops necessary to just decolorize the water sample. (Complete decolorization occurs when all of the iodine in the sample reacts with the vitamin C.)

**Important...**
Be sure to thoroughly mix each drop of vitamin C solution into the water sample before deciding to add another drop.

Page 74

Water Purification—How Much Iodine Is Present?

# How Much Iodine Is Present?

## Leader Guide

2. Prepare the standard vitamin C solution by dissolving 0.500 g (500 mg) vitamin C in 100 mL water.

## Introducing the Activity

Tell campers that, by monitoring the level of chlorine in the water, companies can be sure that enough chlorine is used to kill the bacteria but not so much that it makes people sick. Iodine is substituted for chlorine in this activity because it also kills bacteria, is highly colored, and reacts with vitamin C. Campers will simulate the monitoring of drinking water by using titration to determine how much iodine is present in an unknown water sample. Titration is one way to determine chemical concentrations in water.

## Procedure Notes and Tips

- Inform campers that, as the titration nears its endpoint, a drop of vitamin C can make the povidone-iodine solution go colorless for several seconds, but that continued mixing may bring the color back. Only when the color is completely gone after 10 seconds or so of mixing is the iodine really fully reacted.

- After determining the average number of vitamin C drops required to decolorize the water sample, campers should multiply that average by $1 \times 10^{18}$ to get the number of particles of iodine originally present in their water sample.

**Q1:** Answers will vary, but the following table shows sample answers.

| Water Sample | Average Number of Vitamin C Drops | Approximate Iodine Particles Present |
|---|---|---|
| 5 mL PI/2 L water | 2 drops | $2 \times 10^{18}$ particles |
| 10 mL PI/2 L water | 3 drops | $3 \times 10^{18}$ particles |
| 20 mL PI/2 L water | 6 drops | $6 \times 10^{18}$ particles |
| (optional) 50 mL PI/2 L water | 17 drops | $17 \times 10^{18}$ particles |

---

## Camper Notebook

4. In the data table at the end of this activity, record the number of vitamin C drops required to decolorize your water sample.

5. Repeat steps 2–4 two more times to get trial 2 and trial 3 results. Average your results (to obtain a more accurate number of vitamin C drops required to decolorize your water sample) and record your answer in the data table.

6. Use the average number of drops of vitamin C solution to calculate the approximate amount of iodine particles in your water sample. Here's why and how.

- The decolorization of iodine occurs with one particle of vitamin C reacting with one particle of iodine. A low number of drops indicates little iodine is present, while a high number of drops means more iodine is present.
- Finding the approximate number of iodine particles in your water sample requires some adjustment. Fill in the data table based on the following information.

approximate iodine particles present = ☐* drops of vitamin C $\times 10^{18}$

*Use the average value rather than one trial's value.

If you are curious, your leader can explain where this adjustment factor comes from.

- In a similar way, water companies can monitor chlorine by measuring how many drops of a solution are needed to react with all the chlorine in a water sample.

**Q1:** What is the approximate number of iodine particles in your sample?

| Identity of Water Sample | Number of Vitamin C Drops | | | | Approximate Iodine Particles Present |
|---|---|---|---|---|---|
| | Trial 1 | Trial 2 | Trial 3 | Average | |
| A | | | | | $\times 10^{18}$ |
| B | | | | | $\times 10^{18}$ |
| C | | | | | $\times 10^{18}$ |
| D | | | | | $\times 10^{18}$ |

**By the way...**
$1 \times 10^{18}$ is the same as 1 billion billion or 1 billion$^2$. That's 1 followed by 18 zeros!

**FYI...** You will get data for the other water samples from other campers after everyone is done.

# LEADER

## How Much Iodine Is Present?

**Leader Guide**

### Wrap-Up

Have campers share their results with other groups having different water samples. If you'd like, review the calculation with the campers that determines the particles of iodine in each water sample. (For help with the discussion, see Leader Background Information.) Depending on the ages of the campers, you may want to have them research, identify, and discuss different methods of determining chemical concentrations in water (such as gravimetric, electrochemical, spectrophotometry, titration, chromatography, and mass spectrometry).

**Leader Guide**

# Removal of Iodine from Water

> ✋ Suggestions for leading this activity begin on the next page.

Public and municipal water systems commonly add chlorine compounds to the water (chlorination) to remove bacteria and other biological contaminants. The amount of chlorine used is critical. Enough chlorine must be added to kill bacteria on contact and provide residual (left over) chlorine to help protect against any bacteria introduced into the water as it is delivered through the water delivery pipes. However, the amount of chlorine remaining in the water at the tap must be small enough that it is not harmful to anyone drinking the water. Some home filtration systems contain activated charcoal and promise to remove some or all of the residual chlorine in water. In this activity, the ability of activated charcoal to adsorb (accumulate on its surface) impurities in water is investigated. Since chlorine in water is almost colorless, campers will substitute iodine, which is amber-brown in water, for chlorine because the removal of the iodine color can be visually observed. Iodine is a good substitute for chlorine in this study because it is in the same family as chlorine in the Periodic Table and undergoes similar chemical reactions. (Iodine is used as an antiseptic to topically kill germs.)

During adsorption, many contaminants in water adhere to the surface of the charcoal. These contaminants are removed from the water as the water continues through the charcoal. Adsorption is different than absorption. Adsorption occurs only on the surface of a material, and absorption occurs throughout the material.

By mixing an iodine-water solution (unknown water samples A, B, C, or D) and some activated charcoal, campers will be able to see how well the charcoal acts as a filter. Since the iodine-water solution is an amber-brown color, any color remaining in the water samples after filtration means that some iodine remains in the samples. The darker brown the solution, the more iodine is present. Campers should observe that activated charcoal filtration removes visible iodine at lower initial concentrations, but some iodine remains after filtration at higher initial concentrations. This result demonstrates that charcoal will only remove a specific amount of contaminants. As a result, commercial charcoal filters need to be changed on a regular basis.

Activated charcoal cannot filter out all foreign chemicals. Caution campers not to try this on a contaminated water sample with the intention of drinking the water afterwards.

## Some Leader Background Information

Activated charcoal (charcoal that has been treated with oxygen) removes some water contaminants because the carbon granules provide a surface on which contaminants (represented by the iodine in this activity) can stick. This process, called adsorption, is behind many water purification methods. It's estimated that 1 tablespoon of activated charcoal has a total surface area of about 4,200–28,000 square meters! By comparison, a football field including both end zones has a surface area of about 5,350 square meters.

# LEADER

## Removal of Iodine from Water

### Leader Guide

### Materials for Camper Procedure—See Left

### Materials for Getting Ready

✓ 10% povidone-iodine solution (such as Betadine®)

⚠ *The 10% povidone-iodine solution is available in drugstores as a topical antiseptic microbicide. Iodine itself is toxic and can cause irritation, stains, and burns to skin. However, iodine in the 10% povidone-iodine solution is nonirritating and nonstraining to skin and natural fibers. Be sure campers do not consume any iodine-treated solutions, either before or after activated charcoal treatment.*

✓ 10-mL graduated cylinder

✓ 3 or 4, 2-L graduated beakers or graduated cylinders

### Getting Ready

1. The solutions used in this activity are the same solutions that are used in the activity "How Much Iodine Is Present?" In graduated beakers or cylinders, make the following solutions. These solutions will serve as unknown water samples for the campers. The 2-L amounts will be enough for about 40 campers to test one sample and share results (as outlined in the camper procedure) or for 11–13 campers to test each sample.

   - 5 mL povidone-iodine (PI) diluted to 2 L (2000 mL)
   - 10 mL PI diluted to 2 L
   - 20 mL PI diluted to 2 L
   - (optional) 50 mL PI diluted to 2 L

   ☞ *The charcoal removes less iodine from this more-concentrated iodine solution. Both the initial solution and the charcoal-treated solution are quite dark brown.*

2. Label the water samples A through C or D, but not in consecutive order. Keep an answer key for yourself.

---

# CAMPER

## Removal of Iodine from Water

### Overview

Some people don't want chlorine in their drinking water so they use home filtration systems that promise to remove some or all of the residual chlorine. How do home water filtration systems work? How effective are they at removing chlorine and other substances?

### Materials

✓ water sample prepared by leader

⚠ *The water sample contains iodine from a topical antiseptic. Do not drink the water sample, either before or after filtering.*

✓ 3 clear plastic cups (at least 10-ounce size) or 3, 200-mL beakers

✓ measuring cup, 100 mL graduated cylinder, or graduated beaker

✓ 2 self-stick labels

✓ white paper for a background

✓ 20 g (heaping 1 tablespoon) aquarium charcoal

✓ balance or tablespoon

✓ spoon or stirring rod

✓ filter (such as filter paper or coffee filter) and funnel (such as 60° laboratory funnel or funnel cut from a 2-L plastic soft-drink bottle)

✓ disposable pipet, dropper, or small plastic squeeze bottle with dropper dispenser

✓ vitamin C solution prepared by leader for "How Much Iodine Is Present?"

✓ (optional) pitcher water filter system

### Procedure

1. Measure ⅓ cup (or 75 mL) of your assigned water sample and pour it into a beaker or cup. Label this container "control." Place the container on a white sheet of paper. This sample will be used for the color comparison in step 4.

2. Measure another ⅓ cup (or 75 mL) of your water sample and pour it into another beaker or cup. Add 20 g (or a heaping tablespoon) of aquarium charcoal to the container. Stir the mixture for about 1 minute.

# Removal of Iodine from Water

## Leader Guide

3. See the Leader Guide for "How Much Iodine Is Present" if you need to make more vitamin C solution.

## Introducing the Activity

Tell campers that some water filtration systems use activated charcoal to filter out chlorine and other substances. Since chlorine in water is almost colorless, in this activity we will substitute amber-brown iodine to demonstrate how activated charcoal works. (Iodine is a good substitute for chlorine because it is in the same family as chlorine in the Periodic Table and therefore undergoes similar chemical reactions.)

## Procedure Notes and Tips

- A pitcher water filter system that uses a charcoal filter can be substituted for the activated charcoal and coffee filters. If campers use this pitcher system, the same filter can be used by all campers for all water samples. Try to use a filter that is nearly expired, since you will have to throw the filter away after this activity. Be sure to thoroughly wash out the pitcher after the activity.

- The mixture in step 2 only needs to be stirred for about 1 minute. A short contact time in this activity is okay, particularly since charcoal filters in household water filtration systems work as the water passes through them.

- The two solutions in step 4 must be in the same size and shaped containers in order to visually compare them.

**Q1:** The darker brown the solution, the more iodine is present. A color comparison between the control and the treated sample indicates that the charcoal removes at least some of the iodine from the water sample. Campers should observe that the brown color is darker in the solution before charcoal treatment.

**Q2:** The results depend on the actual charcoal used and on the subjective judgement of the camper. Some campers will report that the charcoal removed all of the iodine from the most dilute solution.

---

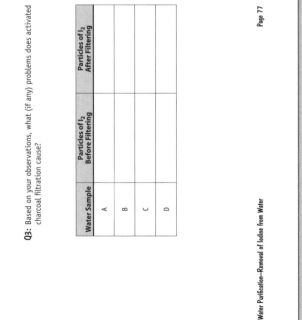

### CAMPER

# Removal of Iodine from Water

**Camper Notebook**

3. Pour the mixture into a filter apparatus to remove the solid charcoal, allowing the liquid to drain into a clean container. Label this container "filtered."

4. Compare the color intensity in the control with the color intensity in the solution treated with charcoal and filtered.

**Q1:** What evidence is there that the charcoal treatment removes some iodine?

5. Use the procedure outlined in the activity called "How Much Iodine Is Present?" to measure how much iodine is in the initial water sample ("control") and how much is left in the water sample after treating it with activated charcoal ("filtered.")

6. In the data table below, enter your results and those of other campers who began with water samples having different initial concentrations of iodine ($I_2$).

**Q2:** At what level of iodine in water does the charcoal filter remove all of the detectable iodine?

**Q3:** Based on your observations, what (if any) problems does activated charcoal filtration cause?

| Water Sample | Particles of $I_2$ Before Filtering | Particles of $I_2$ After Filtering |
|---|---|---|
| A | | |
| B | | |
| C | | |
| D | | |

*Example of filter apparatus* (coffee filter, soft-drink bottle funnel, cup)

# LEADER

## Removal of Iodine from Water

### Leader Guide

**Q3:** Answers may vary, but campers may report that the coffee filter holding the activated charcoal allowed some fine pieces of charcoal into the charcoal-treated water. These pieces would be unacceptable in real drinking water. Commercial charcoal filters use a different, less dusty form of charcoal and a finer filter to minimize this problem.

### Wrap-Up

Discuss adsorption and how activated charcoal is effective in removing contaminants from water. (For help with the discussion, see Leader Background Information.)

## Leader Guide

# What's Really in the Bottle?

👉 Permission is granted for you to copy and distribute this take-home activity for your event. (The master is provided in the Camper Notebook.)

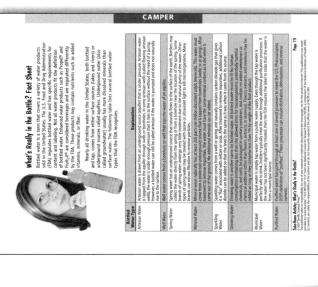

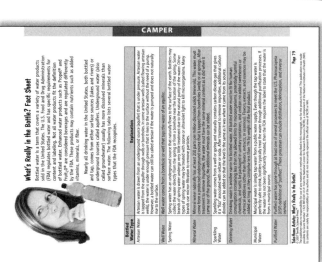

Water Purification—What's Really in the Bottle?

Page 197

## LEADER

# The Amazing Water Maze

👉 Permission is granted for you to copy and distribute this take-home activity for your event. (The master is provided in the Camper Notebook.)

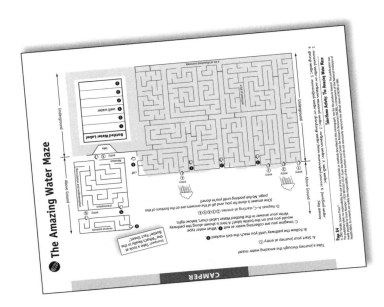

# TOPIC 3

## HEALTHY AIR

### List of Activities

- ✓ Look at One Part per Million ............................................. 200
  *Create a one part per million (1 ppm) solution to learn that 1 ppm is not necessarily an insignificant quantity.*

- ✓ Balloon Challenge ............................................................ 203
  *Explore the fact that air takes up space.*

- ✓ Search for One Part per Million ......................................... 206
  *Discover how easy it is to detect one part per million (1 ppm).*

- ✓ Humidity Detector ............................................................ 209
  *Discover how humidity varies from place to place.*

- ✓ How Much Oxygen Is in Air? ............................................. 212
  *Determine the amount of oxygen in air by observing the oxidation of iron.*

- ✓ Pour a Gas ....................................................................... 216
  *Observe the presence of an invisible gas.*

- ✓ Pour More Gas ................................................................. 219
  *Learn how to detect an invisible gas.*

- ✓ Stirring Indoor Air ............................................................ 222
  *Observe the effects of both tightly sealed and well-ventilated buildings on indoor air pollution.*

- ✓ Trapping Particulates ....................................................... 225
  *Explore the ability of vacuum bags to trap fine particles such as biological contaminants.*

- ✓ Cartesian Diver ................................................................ 228
  *Investigate the compressibility of air.*

- ✓ Take-Home Activity: Growing Mold .................................. 231
  *Learn what conditions help mold to grow.*

Page 199

LEADER

# LEADER

**Leader Guide**

# Look at One Part per Million

 **Suggestions for leading this activity begin on the next page.**

Scientists measure air pollutants using special instruments that are sensitive enough to detect parts per million (ppm) concentrations. In this activity, campers mix food color and water to make a 1 ppm solution. The ease and speed with which they make the 1 ppm solution will help to emphasize that 1 ppm is not necessarily an insignificant quantity.

## Some Leader Background Information

Each dilution of food color is a 10-fold step down in concentration. (That is, each dilution lowers the concentration by a factor of 10.) The first dilution results in a 10% solution (1 in 10), the second dilution results in a 1% solution (1 in 100), and so on. The sixth dilution gives a 1 ppm solution (1 in 1,000,000). At some point (depending on the food color used in the activity), a dilution will probably appear colorless. For example, the color may be clearly present in container 4, very faint in container 5, and not visible at all in container 6. Nevertheless, the food color is still present in all dilutions.

In this activity, campers prepare a 1 ppm solution of a liquid food color. In actuality, the beginning food color is already dissolved in water so the absolute concentration of food color in the end dilution (container 6) is really less than 1 ppm.

# Look at One Part per Million

## Leader Guide

### Materials for Camper Procedure—See Left

### Materials for Getting Ready

✓ white Styrofoam® egg cartons (one carton for every two set ups)
✓ scissors

### Getting Ready

1. Cut whole egg cartons down the middle lengthwise so that you end up with two long carton portions having six depressions side by side.

### Introducing the Activity

You may want to demonstrate that holding the dropper vertically (straight up and down) will give consistent drops.

### Procedure Notes and Tips

- To save on droppers, campers may wash and dry the same ones between dilutions. However, be sure campers rinse and dry the droppers very thoroughly in order to make accurate dilutions.
- Here are the dilutions for each container.

|  | Container | | | | | |
|---|---|---|---|---|---|---|
|  | 1 | 2 | 3 | 4 | 5 | 6 |
| concentration* | 1 part per 10 | 1 part per 100 | 1 part per 1,000 | 1 part per 10,000 | 1 part per 100,000 | 1 part per 1,000,000 (1 ppm) |
| Is color visible? | answer varies | answer varies | answer varies | answer varies | answer varies | answer varies |

* Campers may represent the concentrations in different ways. For example, 1 part food color per 10 parts solution may also be written as 1 part per 10, 1/10, 0.1, and $1\times10^{-1}$.

Page 201

---

# Look at One Part per Million

## Overview

Environmental contaminants, such as pollutants and greenhouse gases, are often measured in parts per million (ppm). One ppm sounds like a pretty tiny amount, but let's see how easy it is to make a 1 ppm concentration.

## Materials

✓ egg carton (prepared by leader) or 6 small white or clear containers (such as plastic cups)
✓ self-stick labels or waterproof marker
✓ piece of white paper (if clear containers are used)
✓ 6 droppers or disposable pipets
✓ water
✓ blue or green liquid food color

## Procedure

1. Number six depressions in the egg carton from 1 to 6. Alternately, number six containers and place them side by side (on white paper if they are clear).

2. Use a dropper to put 9 drops of water into each container. Be sure to hold the dropper vertically (straight up and down) to keep the size of the drops consistent.

3. Use a clean, dry dropper to put 1 drop of food color into container 1. Stir with the dropper to mix well. The concentration of this solution is 1 drop of food color per 10 drops of total solution (1 part per 10). This is recorded for you in the data table at the end of this activity. Record whether the color is still visible at this concentration.

4. Use a clean, dry dropper to transfer 1 drop from solution container 1 to container 2. Stir to mix well. This is 1 part food color per 100 parts of solution. Record this dilution in the data table and whether the color is still visible.

Page 86

# LEADER

## Look at One Part per Million

### Leader Guide

**Q1:** Answers will vary. The dilution at which the color is no longer visible depends on the color of the food color used in the activity. Darker colors will become invisible later than lighter colors.

### Wrap-Up

Ask campers to discuss their results. If time permits, you may want to describe a visual image of one part per million. For example, you can tell campers:

- One part per million can be imagined in several different ways.
  - If you laid peas on the ground in one long line, 1 million peas would stretch for about 4 miles. So, 1 ppm is the ratio of one pea to the single 4-mile line of peas. (By the way, it would take you about 1 hour and 20 minutes to walk the entire length of that 4-mile line of peas.)
  - One million playing cards approximately covers the area of a football field, so 1 ppm is the ratio of one playing card to enough playing cards to cover a football field.
- Can any of you come up with other ways to visualize 1 ppm?

The 1977 film *Powers of 10* explores the relative size of things from the microscopic to the cosmic. The website *powersof10.com* offers many ways to explore this material, including interactive features and the option to purchase the movie.

---

## CAMPER

### Camper Notebook

5. Continue this process for the remaining containers. The last dilution is 1 part food color per 1,000,000 parts of solution, or 1 ppm.

   **Q1:** At what dilution does the solution first appear colorless?

| | \multicolumn{6}{c|}{Container} |
|---|---|---|---|---|---|---|
| | 1 | 2 | 3 | 4 | 5 | 6 |
| concentration | 1 part per 10 | | | | | |
| Is color visible? | | | | | | |

## Leader Guide

# Balloon Challenge

> 👉 **Suggestions for leading this activity begin on the next page.**

When we examine a bottle that does not contain liquid, the word we typically use to describe it is "empty." However, it is actually full of air, a mixture of gases that has its own properties. In this activity, campers will explore one of these properties (namely, that air takes up space).

## Some Leader Background Information

Blowing up a balloon involves forcing additional air into the balloon. The gas particles blown into the balloon hit the inside walls of the balloon, creating enough pressure to force the rubber of the balloon to expand and the balloon to inflate. Pressure also exists outside the balloon (atmospheric pressure). Atmospheric pressure is a result of molecules of gas in the atmosphere pushing on an object.

For the balloon to stay inflated, the pressure inside the balloon must be greater than atmospheric pressure. The pressure inside the balloon must not only counter the atmospheric pressure but also hold the elastic rubber in the inflated position. If the mouth of the inflated balloon is opened (or the balloon pops), the air inside will quickly flow out because gases move from areas of high pressure to areas of lower pressure.

Step 1 of the activity shows that it is impossible to significantly inflate a balloon by mouth when the balloon is inside a bottle. The pressure of the air trapped inside the bottle prevents you from inflating the balloon. In order to blow up the balloon in the bottle, you not only need to blow enough air into the balloon to provide the pressure needed to stretch the rubber of the balloon, but you also have to apply enough pressure to compress the air trapped in the bottle. This compression is needed to make room for the inflated balloon. The activity shows this to be a real challenge.

Even though gases are compressible, it is difficult for most of us to exert enough pressure just by blowing to compress the trapped air very much.

Step 3 shows that a balloon can be blown up in a bottle far more easily if there is a hole in the bottle that allows air initially inside the bottle to escape and make room for the expanding balloon. If you seal the hole after inflating the balloon (as directed in step 5), the balloon will remain inflated even after you remove your mouth. For the balloon to deflate, it is necessary for air particles to occupy the volume previously occupied by the inflated balloon. However, a careful observer may notice that the bottom part of the plastic bottle contracts around the inflated balloon or the balloon very slightly shrinks. The atmospheric pressure exerted on the open balloon is greater than the pressure in the region between the balloon and the bottle, which is enough to keep the balloon inflated except for this slight shrinkage. The balloon will deflate only when the hole in the side of the bottle is opened again and air moves into the bottle to replace the air coming out of the bottle from the balloon.

# LEADER

## Balloon Challenge
### Leader Guide

### Materials for Camper Procedure—See Left

### Materials for Getting Ready

✓ empty and clean 1-L or 2-L soft-drink bottle (one per camper)

☞ Campers will use another empty and clean 1-L or 2-L soft-drink bottle in step 1. No extra preparation is necessary for that bottle.

✓ sharp object such as pushpin, thumbtack, or nail

### Getting Ready

1. Prepare one bottle per camper by poking a hole on the side of the bottle near the bottom. (This bottle will be used in steps 3–5.)

### Introducing the Activity

Blow up a balloon in front of the group and discuss how the balloon expands as air pushes into it during the inflating process. (For help with the discussion, see Leader Background Information.) Tie off the balloon and gently squeeze it to show that the balloon cannot be completely flattened without popping it.

### Procedure Notes and Tips

- Make sure each camper uses the same-sized bottles throughout the activity (either 1 L or 2 L).
- For health reasons, make sure each camper has his or her own balloon.
- Give campers the bottles with the holes (prepared in Getting Ready) after they have completed steps 1 and 2.

**Q1:** The balloon will not blow up very much.

**Q2:** When the balloon is out of the bottle, it can expand when air is blown into it.

**Q3:** Some campers may realize that, in step 1, the air trapped in the bottle stops the balloon in the bottle from expanding very much. Others may

Page 204

---

# CAMPER

## Balloon Challenge

### Overview

Is an empty bottle really empty? We usually describe a bottle that does not contain a liquid or solid as being empty, as if the air inside does not exist. But air is present and has properties of its own. In this activity, you will examine one of these properties.

### Materials

✓ balloon
✓ empty and clean 1-L or 2-L soft-drink bottle
✓ soft-drink bottle prepared by leader

### Procedure

1. Push a deflated balloon into the first soft-drink bottle and stretch the open end of the balloon back over the mouth of the bottle as shown. Now blow into your balloon.

Q1: What happens?

2. Take your balloon out of the bottle and blow into the balloon.

Q2: What happens?

Q3: Can you explain the difference between what happens in step 1 and what happens in step 2?

3. Push your deflated balloon into another bottle (one that your leader gives you) just like you did in step 1. Now blow into your balloon.

Page 88

Healthy Air–Balloon Challenge

# Balloon Challenge

## Leader Guide

realize that the balloon in the bottle stops expanding when the air pressure inside the balloon and inside the bottle equalize.

**Q4:** The balloon in the bottle inflates because the air in the bottle can escape through the hole in the side of the bottle. Some campers may realize that steps 1–3 demonstrate that air takes up space. Since the air can't escape from the bottle in step 1, there is little room for the balloon in the bottle to expand very much. Once the air can escape from the bottle in step 3, there is enough space for the balloon in the bottle to expand.

**Q5:** After placing a finger on the hole, attempting to inflate the balloon is difficult. The finger closes the hole and does not allow air to escape from the bottle as air is forced into the balloon. The bottle and balloon behave similarly to the bottle and balloon in step 1.

**Q6:** Once the balloon is inflated inside the bottle, placing a finger on the hole allows the balloon to remain inflated inside the bottle even when the person's mouth is removed. The finger on the hole prevents air from entering the bottle, so outside air pressure can not deflate the balloon.

## Wrap-Up

Discuss the answers to the questions as well as camper observations. Explain why the balloon behaves as it does in each situation. (For help with the discussion, see Leader Background Information.)

---

## Camper Notebook

**Q4:** What happens? Can you explain why this happens? What do the results of steps 1–3 tell you about what's in the bottle?

4. With the setup used in step 3, hold your finger over the hole on the side of the bottle and then blow into your balloon.

**Q5:** What happens and why?

5. With the setup used in step 3, leave the bottle's hole open and blow into your balloon. With your mouth still over your balloon's opening, place a finger tightly over the bottle's hole. Now remove your mouth but keep your finger over the hole.

**Q6:** What happens and why?

# LEADER

## Leader Guide

# Search for One Part per Million

> Suggestions for leading this activity begin on the next page.

Scientists measure air pollutants using specialized equipment. This equipment takes advantage of unique characteristics of the substances scientists are looking for. The equipment often measures small concentrations of materials in parts per million (ppm). While this may sound like an impossibly tiny concentration, campers will discover that 1 ppm isn't really all that small. This is why some air pollutants detected in parts per million can have dramatic effects on people sensitive to those pollutants.

## Some Leader Background Information

With the jar in step 1, campers try to find one dark-colored particle in about a million light-colored grains of salt. The most effective search method for finding the particle is to slowly roll the jar on its side. Generally the particle can be found within a few minutes.

Step 2 of the activity with the glitter and water shows campers another example of what 1 ppm looks like. One part per million can be understood as 1 mg (or 0.001 g) of a substance in 1 kg (or 1000 g). Since an average piece of glitter weighs 1 mg and 1 L of water weighs 1 kg, a mixture that has 1 mg of a substance (for example, one piece of glitter) in 1 L of water is about 1 ppm.

In this activity, campers use vision to look for the "target" substance. Its contrasting color makes the particle unique when compared to the white salt grains in the jar or water in the bottle. In the same way, environmental scientists can find a "one in a million" very easily by using specialized equipment designed to take advantage of a unique characteristic of the "target" substance. For example, a magnet will find a steel bead among 999,999 nonmetallic beads without difficulty. There are many other ways to isolate a target substance. In all cases, the method concentrates on a property (unique characteristic) of the target substance that makes it different than other substances, therefore making it easy to isolate.

**Unique characteristic...**
If a contrasting bead were larger than the other beads, it could easily be sifted out with a strainer that had contrasting holes smaller than the other beads. bead but bigger than the other beads.

Healthy Air—Search for One Part per Million

# Search for One Part per Million

## Leader Guide

### Materials for Camper Procedure—See Left

### Materials for Getting Ready

- ✓ dry, clear jar with a capacity of about 16 ounces (500 mL) and lid (one per camper)
- ✓ salt
- ✓ dark-colored particle (such as grain of colored sugar, grain of colored sand, or very dark flake of black pepper)
- ✓ ½-cup dry measuring cup or other volume measure
- ✓ self-stick labels
- ✓ clear 1-L soft-drink bottle with cap (one per camper or group)
- ✓ glitter
- ✓ water
- ✓ (optional) static-reducing dryer sheets or spray

### Getting Ready

1. Prepare the 1 ppm salt and pepper jars.

   a. If the clear jars are plastic, you can reduce static cling by rubbing the inside of the jars with a dryer sheet. Static-reducing spray also works, but be sure the insides are completely dry before starting the next step.

   b. Put ½ cup (125 mL) salt in each dry jar.

   c. Add one dark-colored particle to each jar.

   d. Tightly close the lids and shake the jars. Label the jars "1 ppm."

---

## Search for One Part per Million

### Overview

Scientists measure air pollutants using special instruments that are sensitive enough to detect part-per-million (ppm) concentrations. How easy is it to detect just 1 ppm without these special instruments?

### Materials

- ✓ jar containing one dark-colored particle in white salt (prepared by leader)
- ✓ watch or clock with second hand
- ✓ bottle of water containing one piece of glitter in the water at a concentration of 1 ppm.

### Procedure

1. Time and record at left how long it takes you to find the dark-colored particle that is "one in a million." If you'd like, shake the jar and time your search several more times.

2. Now look at the bottom of the bottle of water. There is one piece of glitter in the water at a concentration of 1 ppm.

**Q1:** Does the piece of glitter disappear in the water or is it easily visible? Why?

**Q2:** How can the glitter be removed?

| I found the particle in how much time? | | | |
|---|---|---|---|
| | | | |

### FYI...

- 200 grains of salt weigh about 0.033 g.
- 1,000 grains of salt weigh 0.165 g.
- 1,000,000 grains weigh 165 g.
- 165 g of salt has a volume of ½ cup (about 125 mL).

# LEADER

## Search for One Part per Million

### Leader Guide

2. Prepare the 1 ppm glitter and water bottles.

    a. For each group of campers, put one piece of glitter into a soft-drink bottle.

    b. Fill the bottles with 1 L water and tightly seal them. Label the bottles "1 ppm."

### Procedure Notes and Tips

- The most effective way to find the dark-colored particle is to slowly roll the jar on its side.

**Q1:** Since the glitter does not dissolve in the water, it remains easily visible. Some campers may relate that glitter is different than salt and sugar, which dissolve (disappear) in water.

**Q2:** One way to remove the glitter would be to filter the glitter out. Another would be to allow it to settle to the bottom of the bottle and carefully pour off the water. Campers will probably come up with additional ways. In all cases, the method should target a property of the glitter that makes it different from water (such as solubility or density) and therefore makes is easily isolated from water.

### Wrap-Up

Ask campers to share their observations and answers to the questions. Discuss how the activity relates to specialized equipment that environmental scientists use to detect low levels of chemicals in air and water. (For help with the discussion, see Leader Background Information.) If you'd like, you or the campers may research and discuss some analytical methods used to test for pollution in air and/or water.

# Humidity Detector

**Leader Guide**

> 👆 Suggestions for leading this activity begin on the next page.

Dry air is composed of about 78% nitrogen, 21% oxygen, and 1% argon. However, water is another gas that is present in the air. Too much moisture (humidity) in the air promotes the growth of mold and mildew. Too little can cause dry eyes and dry skin. Although relative humidity can be quantitatively measured with specialized instruments called hygrometers, the humidity detector in this activity can give campers an idea of how the humidity varies from location to location.

## Some Leader Background Information

Air can only hold a given amount of water at a given temperature. Air that is holding as much moisture as possible is said to be saturated, or at 100% (relative) humidity. Air that is holding half as much moisture as possible at a given temperature is at 50% relative humidity. Warm air can hold more moisture than cold air, which is why high relative humidity in the summer is more unpleasant than the same relative humidity in the winter.

The flower colors observed in this activity are a result of the loss or gain of water. Anhydrous (without water) cobalt chloride ($CoCl_2$) is blue while cobalt chloride hexahydrate (cobalt chloride with six water molecules) is pink. Blue cobalt chloride is observed when little or no water is present (for example, in very dry air). The pink color is observed when the amount of water present is great (for example, in humid air or aqueous solutions). A purple color represents a combination of the anhydrous and hydrated forms, indicating a moderate level of water.

The chemical reaction can be represented as follows:

$$CoCl_2 + 6H_2O \leftrightarrows CoCl_2 \cdot 6H_2O$$
cobalt chloride (blue) + water ⇌ cobalt chloride hexahydrate (pink)

Campers learn from misting the flower with water that high humidity turns the flower pink. When the dry flower is placed in different locations, a pink color indicates high moisture, purple indicates a moderate level of moisture, and blue indicates dry conditions.

# LEADER

## Humidity Detector

# Leader Guide

## Materials for Camper Procedure—See Left

### Supply Information

- Chenille stems are available in craft and hobby stores.
- Cobalt chloride hexahydrate is available from chemical supply companies.

### Materials for Getting Ready

✓ 10 g (about 2 teaspoons) cobalt chloride hexahydrate ($CoCl_2 \cdot 6H_2O$)
  *This amount of cobalt chloride hexahydrate should make enough solution to dip 30–40 flowers. Adjust the amount of solution you make as needed.*

✓ balance or teaspoon

✓ graduated beaker or liquid measuring cup

✓ one or more small shallow containers

### Getting Ready

⚠ *Solid cobalt chloride hexahydrate is highly toxic. Do not inhale or take internally. Wear goggles when handling the solid and the solution. Avoid contact with skin and eyes. Wash skin and eyes well if contact occurs.*

1. Prepare a 10% cobalt chloride solution by dissolving 10 g (about 2 teaspoons) cobalt chloride hexahydrate in 100 mL (½ cup) water.

2. Just before camp, put on goggles and carefully pour the solution in one or more small shallow containers so campers can dip their flowers.

### Introducing the Activity

Emphasize to campers that the cobalt chloride solution is harmful if taken internally. They must not taste it. They should wear goggles when handling the solution. They must wash their skin well if skin contact occurs.

Page 210

---

# CAMPER

## Humidity Detector

### Overview

Dry air is mainly nitrogen, oxygen, and argon. Water vapor is another gas that is present in the air. Air that contains too much moisture may promote mold and mildew growth, which can cause health problems. Air that is too dry may cause uncomfortably dry eyes and skin. Is it possible to detect the relative amount of moisture in the air by a color change? Find out by making a humidity detector.

### Materials

✓ round white filter paper about 8 inches (20 cm) in diameter
  *If a cone-shaped or basket-type coffee filter is used, not as many folds are needed to create the wedge shape in step 1.*

✓ green pipe cleaner or chenille stem about 12 inches (30 cm) in length

✓ 2 green tissue paper rectangles about 4 inches × 6 inches (10 cm × 15 cm)

✓ scissors

✓ 10% aqueous cobalt chloride solution prepared by leader
⚠ *Cobalt chloride is harmful if taken internally. Do not taste the cobalt chloride solution. Wash skin well if contact occurs. Wear goggles when handling the solution.*

✓ shallow container

✓ place to hang humidity detector to dry (such as clothes drying rack or clothesline)

✓ clothespin

✓ hair dryer

✓ fine-mist sprayer bottle filled with water

✓ goggles

### Procedure

1. Fold the round filter paper in half about three times to make a wedge shape. Fold the wedge about 1 inch (3 cm) from the pointed end. Bend the center of the pipe cleaner over the fold as shown and twist the pipe cleaner tightly three times.

wedge of folded filter paper
pipe cleaner or chenille

Page 91

# Humidity Detector

## Leader Guide

### Procedure Notes and Tips

- Campers should only dip the outside of their flowers. Since campers dry the flowers after dipping them, a small application of the solution is preferred.
- Unused cobalt chloride solution may be stored for future use. (Be sure the solution is well labeled and stored safely away from foods and drinks.)

**Q1:** The cobalt chloride solution is pink when it is first applied to the filter paper.

**Q2:** Once the paper dries, the cobalt chloride may remain pink if the humidity is high. If the humidity is low, the compound will turn blue.

### Wrap-Up

Discuss what the different flower colors mean. (For help with the discussion, see Leader Background Information.) Ask campers to share what they discovered during step 6 when they put their flowers in different locations. Review how too much moisture or too little moisture may cause certain health problems.

---

## Camper Notebook

tissue paper leaves

2. Place the center of the tissue paper rectangles into the pipe cleaner and continue twisting the stem to the end as shown.

3. Put on goggles. Hold the "flower" upside down and dip the filter paper briefly into the 10% cobalt chloride solution prepared by your leader.

4. Hang the flower upside down by the stem and dry it with a hair dryer.

**Q1:** What color is the cobalt chloride when the cobalt chloride is in solution and first applied to the filter paper?

**Q2:** What color is the cobalt chloride when the paper is dry?

5. Open the blossom of the flower, spray the flower with water using a mist sprayer, and observe what happens. Dry the flower again like you did in step 4.

6. Place the flower in different locations, such as near a heating vent, ceiling, and open window. Try putting the flower in places such as a closet, basement, attic, and crawl space. In the data table, record each location and the flower's color at that location. Based on what you observed about the flower's color in steps 4 and 5, decide and record the relative amount of moisture in each place (for example, high moisture, moderate moisture, dry).

⚠ *Be sure to keep the flower away from small children and pets.*

| Location | Flower Color | Relative Amount of Moisture |
|---|---|---|
|  |  |  |
|  |  |  |
|  |  |  |
|  |  |  |
|  |  |  |
|  |  |  |
|  |  |  |

# LEADER

## Leader Guide

# How Much Oxygen Is in Air?

> ✋ **Suggestions for leading this activity begin on the next page.**

Oxygen is essential for living things. The air we breathe normally consists of about 21% oxygen. However, unusual circumstances (enclosed spaces, for example) can cause the oxygen level to change. In this activity, campers measure the decrease in air bubble volume while steel wool rusts in an inverted test tube. They use the difference in their measurements from start to finish to calculate how much oxygen was present in the initial bubble.

## Some Leader Background Information

We have all probably seen rust and can recognize it by its characteristic red-brown color, flaky texture, and the pitting it leaves behind on the original metal. We may even know that rusting is the oxidation of iron. However, most of our experience with this process concentrates on the corrosion of iron and the weakening of the metal in the presence of water. We may not realize that oxygen is the other key reactant needed for the reaction to occur. This activity provides evidence of the consumption of oxygen during the rusting process.

Campers place steel wool in a test tube and invert the test tube in water. In the presence of water, iron in the steel wool combines with oxygen in the trapped air and iron(III) oxide (rust) is formed. Oxygen in the trapped air bubble is consumed in the reaction, so the volume of the bubble decreases and more water flows into the test tube. The tremendous surface area of the steel wool allows the depletion of oxygen to occur rather quickly in the system. Also, campers presoak the steel wool in acetic acid to provide an electrolytic solution which improves ion transfer (compared to using just distilled water in the system).

If a chemical equation is used to describe the rusting process, the process can be described as iron + oxygen → rust (or iron oxide). If chemical symbols are used, then $4Fe + 3O_2 \rightarrow 2Fe_2O_3$ is one way to correctly represent the rusting process. (The elements nitrogen and oxygen in air are mostly present as diatomic gases. Nitrogen exists as two nitrogen atoms bonded together in a nitrogen molecule, $N_2$. Oxygen also most commonly exists as two oxygen atoms bonded together in an oxygen molecule, $O_2$.)

# How Much Oxygen Is in Air?

## Leader Guide

## Materials for Camper Procedure—See Left

## Materials for Getting Ready

✓ graduated beaker capable of holding 500 mL
✓ 300 mL vinegar
✓ water
✓ (optional) materials to make a dry air jar

- clean and dry 1 gallon (4 L) transparent jar with lid
- static-reducing dryer sheet or spray
- cake decoration nonpareils

☞ *Nonpareils are the tiny, ball-shaped sprinkles put on candy and cakes. You can find nonpareils in stores selling cake decorating supplies. Nonpareils are sold by weight (net ounces). Because a net ounce of nonpareils is roughly equal to a fluid ounce of nonpareils, you will measure them with a liquid measuring cup in this activity. You will need 4 different contrasting colors, including 86 net ounces (2.4 kg) of one color, 23 net ounces (650 g) of another color, and about 1 net ounce (28 g) of a third and fourth color.*

- liquid measuring cup
- measuring spoons
- self-stick label

## Getting Ready

1. Prepare 0.5 M acetic acid by measuring 300 mL vinegar in a beaker capable of holding 500 mL. Fill the beaker to the 500-mL mark with water.

2. If you'd like, you can make a jar representing the main components of indoor air to show campers during Wrap-Up.

   - If the jar is plastic, rub the inside surface of the jar with a static-reducing dryer sheet to reduce the static clinging of nonpareils to the jar surface. As an alternative, spray some static-reducing spray into the jar and allow the jar to dry thoroughly.

---

## CAMPER

# How Much Oxygen Is in Air?

### Overview

Air is made up of many gases. Oxygen, one of the gases in air, is essential for living things. How much of the air is oxygen? In this activity, you will perform a chemical reaction and take measurements to find out.

### Materials

✓ about 20 mL 0.5 M acetic acid (prepared by leader)
✓ 50-mL beaker
✓ 1 g fine or medium-fine steel wool
✓ balance capable of measuring 0.1 g
✓ paper towels
✓ test tube about 15 cm (6 inches) long
✓ metric ruler at least 15 cm (6 inches) long
✓ tape that will stick when submerged in water
✓ 400-mL beaker or a cut-off 1-L clear soft-drink bottle
✓ water
✓ stirring rod
✓ watch or clock with second hand

### Procedure

1. Pour about 20 mL of the 0.5 M acetic acid into a 50-mL beaker. (This will be used to activate the steel wool's surface.)

2. Immerse about 1 g steel wool in the acetic acid solution for about 1 minute. Remove the steel wool and press the excess liquid out with paper towels. (Be careful because the steel wool is sharp.) Rinse your hands with water after handling the acetic acid solution.

3. Tape a metric ruler to the side of a test tube so that the zero mark is at the mouth (open end) of the test tube as shown. Measure the length of the test tube. To compensate for the tube's rounded bottom, measure to only halfway up the bottom's curve as shown. Record the length of the test tube in the data table on the next page. Be sure to include the units in your measurement.

# LEADER

## How Much Oxygen Is in Air?

### Leader Guide

- To represent nitrogen (780,000 ppm), add 86 fluid ounces (2.5 L) of one color nonpareils to the jar.
- To represent oxygen (209,000 ppm), add 23 fluid ounces (680 mL) of a second color of nonpareils to the jar.
- To represent argon (9,000 ppm), add 2 tablespoons (about 30 mL) of a third color of nonpareils to the jar.
- To represent carbon dioxide (350 ppm), add a scant ¼ teaspoon (1.2 mL) of a fourth color of nonpareils.
- Label the jar with a color key that identifies the color that represents each element.
- Tightly screw on the jar's lid. Gently shake the jar to evenly distribute the nonpareils.

### Introducing the Activity

Warn campers that the steel wool is sharp. Remind campers to wash their hands after soaking the steel wool and after putting the steel wool into the test tube, since 0.5 M acetic acid is a dilute weak acid.

### Procedure Notes and Tips

- Oxygen in the test tube's air bubble reacts with iron in the steel wool, raising the water level inside the tube.
- After the first 9–15 minutes, campers can begin drawing their graphs in step 8 while waiting to take each of the next readings.

**Q1:** The process occurring in the test tube is a fast form of rusting.

---

# CAMPER

## How Much Oxygen Is in Air?

**Camper Notebook**

4. Fill a 400-mL beaker about ⅔ full of water.

5. Carefully pull the steel wool apart to increase its surface area. (Be careful because steel wool is sharp.) Measure the mass of the steel wool and record the information in the data table. Use a stirring rod to insert the steel wool so that it stays at least halfway up the test tube. Wash your hands.

6. Quickly turn the test tube over and place it into the beaker so that the test tube's mouth stays below water level. Record the starting time when the test tube is inserted into the water. Measure and record the water level in the test tube at the starting time. When taking the water level readings, keep the test tube inverted in the water but line up the water level inside the test tube with the water level outside the test tube.

7. Every 3 minutes, record the time and the water level in the test tube. After 15 minutes, the readings can be taken every 5 minutes. Take the last reading at 40 minutes (if time permits).

| Length of test tube: | |
|---|---|
| Mass of steel wool before the reaction: | |
| **Time** | **Water Level in Test Tube** |
| (starting time) | |
| (after 3 minutes) | |
| (after 6 minutes) | |
| (after 9 minutes) | |
| (after 12 minutes) | |
| (after 15 minutes) | |
| (after 20 minutes) | |
| (after 25 minutes) | |
| (after 30 minutes) | |
| (after 35 minutes) | |
| (after 40 minutes) | |

**Q1:** What everyday process is happening in the test tube?

# How Much Oxygen Is in Air?

## Leader Guide

**Q2:** Not all of the steel wool is reacted in the test tube. Some campers may realize that oxygen in the test tube's trapped air is used up in the rusting process.

**Q3:** The steel wool is gaining mass from the oxygen that is combining with the iron atoms to form rust. Since we know the initial mass of the steel wool, we can dry what is left of the steel wool at the end of the experiment, measure its mass, and compare the two masses.

**Q4:** Answers will vary. Some campers may realize that measuring the steel wool's mass after it rusts may reveal a mass greater than it actually is due to water drops on the steel wool. However, if water is removed by drying with paper towels, some rust may also be removed.

## Wrap-Up

Ask campers to share their results. Discuss the rusting process. (For help with the discussion, see Leader Background Information.) Reveal to campers that most of air (about 78%) is nitrogen. Oxygen is about 21% of air. About another 1% of air is argon gas. Other gases are present in significantly smaller amounts. If you'd like, you can show campers a jar that represents the main components of indoor air. (See step 2 of Getting Ready for how to prepare the jar.)

## CAMPER

**Camper Notebook**

**Q2:** Is all of the steel wool reacted in the test tube? If not, what other reactant may be used up in the test tube?

**Q3:** Do you think that the steel wool is gaining mass or losing mass? How could you check?

**Q4:** What else do you need to consider when evaluating the steel wool's mass?

8. Plot a graph showing water level in the test tube versus time. Be sure to label each axis and include units.

**Water Level Versus Time**

9. Do this calculation to determine the percentage of oxygen in air:

$$\frac{\text{final water level in test tube}}{\text{test tube length}} \times 100 = \% \text{ oxygen}$$

Page 95

Healthy Air—How Much Oxygen Is in Air?

# LEADER

## Pour a Gas

> ✋ **Suggestions for leading this activity begin on the next page.**

Carbon dioxide ($CO_2$) is a colorless, odorless gas given off by fossil fuels during combustion and by animals (including human beings) during respiration. Plants use carbon dioxide and water to grow and produce carbohydrates. Carbon dioxide exists in the atmosphere at a concentration of about 360 parts per million (ppm), up from about 280 ppm 250 years ago. In this activity, campers make this gas (without knowing what it is) and use it to extinguish a burning candle.

## Some Leader Background Information

Carbon dioxide is a greenhouse gas that is released when carbon-containing fossil fuels (such as natural gas, coal, and oil) are burned. Steadily increasing levels of carbon dioxide have been implicated as a factor in global warming. Carbon dioxide has a relatively low toxicity (TLV), but can cause unique indoor air quality problems. Since carbon dioxide is about 1.5 times more dense than oxygen, it will sink to the lowest levels it can. In enclosed spaces, too much carbon dioxide and too little oxygen can cause people and animals to suffocate.

A disaster caused by carbon dioxide occurred in the Lake Nyos region of western Cameroon in 1986. A cloudy mixture of carbon dioxide and water droplets suddenly erupted from Lake Nyos. The carbon dioxide, being denser than air, flowed into a nearby valley where it covered a village, displacing the oxygen and killing over 1,700 people. Thousands of cattle and many more animals also suffocated due to lack of oxygen. Most scientists believe that this tragedy occurred due to thermal inversion, when the lower layers of this deep lake were rapidly brought up to the surface by heavy rainfall. Today, large pipes in the lake allow dissolved carbon dioxide to slowly and safely bubble out into the air to prevent future violent explosions.

## Leader Guide

Carbon dioxide is a nonflammable gas that does not support combustion. When campers mix baking soda with vinegar, the reaction releases carbon dioxide that cannot be seen or smelled. Since carbon dioxide is heavier than air, it can be poured out of the bottle and into the foil cup almost like a liquid. When the carbon dioxide is poured into the foil cup, the air in the cup is pushed up and out by the denser carbon dioxide gas. Carbon dioxide collects around the flame, the flame loses its oxygen supply, and the candle goes out.

The chemical reaction between vinegar and baking soda can be written as:

$$CH_3CO_2H(aq) + Na^+HCO_3^- \rightarrow CH_3CO_2^-Na^+ + H_2CO_3(aq)$$

$$H_2CO_3(aq) \leftrightarrows H_2O(l) + CO_2(gas)$$

# Pour a Gas

## Leader Guide

### Materials for Camper Procedure—See Left

### Introducing the Activity

Emphasize to campers that there is a danger of fire even with small birthday candles. The work area must be clear of combustible materials. Loose clothing is discouraged, and long hair must be tied back.

### Procedure Notes and Tips

- Be sure to supervise campers around flames.
- If the candle burns too low to light again comfortably, put in a new one.
- If the candle does not go out in step 6, have campers add more vinegar and baking soda to their bottles and try again.
- When campers are finished, have them pour the bottle's contents down the drain and rinse out the bottle.

**Q1:** Answers will vary. Some campers will think that the gas will put out the candle and others will think that it won't.

### Design your own...

As an extension to the activity, campers can design their own experiments to determine if other household materials can produce carbon dioxide. For example, try:
- Kool-Aid® (citric acid)
- baking powder

---

## Pour a Gas

### Overview

The concentration of a certain colorless, odorless gas in our atmosphere has increased from about 280 parts per million (ppm) in the year 1750 to about 360 ppm in 2000. What is this gas? The following activity shows ways to detect its presence.

### Materials

- birthday candle
- scissors
- foil cupcake liner (discard paper separators)
- aluminum foil or clay
- graduated beaker or liquid measuring cup
- at least ⅓ cup (75 mL) vinegar
- 2-L plastic soft-drink bottle
- matches (for adult use only)

⚠ There is a danger of fire even with small birthday candles. The work area must be clear of combustible materials. Loose clothing is discouraged and long hair must be tied back.

- funnel (a cone made of paper will also work)
- at least 1 teaspoon (5 mL) baking soda
- teaspoon or other volume measure

### Procedure

1. Cut a birthday candle a little bit shorter than the height of a foil cupcake liner. Use a bit of foil or clay to make a holder so the candle stands in the center of the foil cup. The top of the candle's wick should be below the rim of the foil cup.

2. Pour about ⅓ cup (75 mL) vinegar into a soft-drink bottle. Although the bottle now contains a little vinegar, it is mostly filled with the gases that make up air.

**Q1:** Do you think the gas in the bottle will extinguish (put out) a candle flame?

# Pour a Gas

## Leader Guide

**Q2:** The candle does not go out.

**Q3:** The candle does not go out.

**Q4:** The mixture of vinegar and baking soda fizzes. Some campers may realize that the fizzing indicates a gas is being formed.

**Q5:** The candle goes out. Some campers may realize that the gas from the bottle extinguishes the flame because it is heavier than ordinary air. Thus, it displaces the air and deprives the flame of oxygen.

**Q6:** Some campers may say that, if their house was filled with this gas, everyone could suffocate (including pets) due to the lack of oxygen.

## Wrap-Up

Tell campers that mixing vinegar with baking soda makes carbon dioxide gas. Ask campers to list the properties of carbon dioxide that they discovered by doing the activity. Campers should come up with the fact that carbon dioxide is denser than oxygen (they were able to pour it out of the bottle and into the foil cup almost like a liquid) and that carbon dioxide does not support combustion (the candle's flame went out).

---

## CAMPER

### Camper Notebook

3. Light the candle. Slowly tip the bottle as if pouring the gas into the foil cup. Stop tipping before any of the liquid comes out of the bottle.

   **Q2:** Does the candle go out? If so, why?

4. Tip more to pour a few drops of vinegar into the foil cup around the candle. (Do not let the vinegar drops hit the candle flame itself.)

   **Q3:** Does the candle go out? If so, why?

5. Using a funnel or paper cone, add about 1 teaspoon (5 mL) baking soda into the bottle containing the vinegar. Swirl the bottle to make sure the liquid and powder mix well.

   **Q4:** What happens when baking soda is added to vinegar?

6. When the reaction dies down, try tipping the bottle over the foil cup around the lit candle just as you did in step 3. (It may be necessary to keep pouring until a few drops of liquid fall into the foil cup. It may also help to squeeze the bottle to force out some of the gas.)

   **Q5:** Does this gas extinguish the flame? Based on your observations, what did you learn about the gas formed by combining vinegar and baking soda?

   **Q6:** What do you think would happen if your house was completely filled with this colorless, odorless gas instead of air?

## Leader Guide

# Pour More Gas

> ✋ **Suggestions for leading this activity begin on the next page.**

In "Pour a Gas," campers make odorless, colorless carbon dioxide gas and use it to put out a candle flame. Carbon dioxide is denser than air and does not support combustion. In this activity, campers use blue litmus paper to visually detect the presence of carbon dioxide in water.

## Some Leader Background Information

The pH scale is used to designate whether a substance is an acid or base.

$$0 \underset{acid}{\rule{2cm}{0.4pt}} 7 \underset{base}{\rule{2cm}{0.4pt}} 14$$

Lemon juice, for example, is acidic and has a pH of about 4. If you have ever tasted lemon juice, eaten a grapefruit, or accidently tasted spoiled milk, you have experienced the sour taste of acids. On the other hand, soaps are bases and typically have a pH of about 10–11. Bases are known for their bitter taste. If you have ever gotten soap in your mouth or tasted the bitterness of a pill, you have experienced the bitter taste of bases. Bases also feel slippery like soap.

During the first part of this activity, campers dip blue litmus papers into water and club soda and record their observations. Blue litmus paper turns red when dipped into an acidic liquid and stays blue when dipped into a liquid that is not acidic. Plain water is neutral or nearly neutral and therefore does not change the color of blue litmus paper. Club soda is a commercial solution of carbon dioxide in water. Club soda is acidic and therefore turns blue litmus paper red.

After discovering that water is not acidic and club soda is acidic, campers dissolve carbon dioxide into water in step 5. When vinegar is mixed with baking soda, these chemicals react to produce carbon dioxide ($CO_2$) gas. (See a representation of the chemical reactions in Background of "Pour a Gas.") Carbon dioxide gas can be dissolved in water just like salt and sugar.

$$CO_2 + H_2O \rightarrow H_2CO_3$$

When carbon dioxide gas is dissolved in water, the water becomes acidic and blue litmus paper dipped in this water turns red. This weak solution of carbonic acid ($H_2CO_3$) has a tart, sour taste.

# LEADER

## Pour More Gas

### Leader Guide

### Materials for Camper Procedure—See Left

### Supply Information

- Blue litmus paper is available from chemical supply companies.

### Introducing the Activity

Discuss a few general facts about pH, acids, and bases, including the fact that acids typically taste sour. (For help with the discussion, see Leader Background Information.) Remind students that in "Pour a Gas" they determined that carbon dioxide gas is denser than air. Tell campers that litmus is a dye obtained from lichen that is one color (red) in acid solutions and another color (blue) in base (alkali) solutions. Blue litmus paper turns red when dipped in acidic solutions and stays blue when dipped is solutions that are not acidic.

### Procedure Notes and Tips

- Technically, for campers to confirm that water is near neutral and not basic, they would also have to test the water using red litmus paper. (Red litmus paper turns blue when dipped in solutions that are basic and stays red when dipped in solutions that are not basic.) If campers tested the water in step 1 with both blue and red litmus papers, the blue litmus paper would stay blue (indicating that the water is not acidic) and the red litmus paper would stay red (indicating that the water is not basic). These tests would reveal that water is near neutral. Natural and municipal water may deviate slightly from neutrality and therefore give different litmus results.

**Q1:** Campers should observe that the blue litmus paper does not change color, indicating that the water is not acidic.

**Q2:** The blue litmus paper turns red, indicating that club soda is acidic.

Page 220

---

# CAMPER

## Pour More Gas

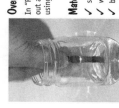

### Overview

In "Pour a Gas," you made a colorless, odorless gas and used it to put out a candle. In this activity, you will try to visually detect the gas using litmus paper.

### Materials

- ✓ small jar with a lid
- ✓ water
- ✓ blue litmus paper
- ✓ club soda
- ✓ small cup
- ✓ about ⅓ cup (75 mL) vinegar
- ✓ graduated beaker or liquid measuring cup
- ✓ 1-L or 2-L plastic soft-drink bottle
- ✓ about 1 teaspoon (5 mL) baking soda
- ✓ teaspoon or other volume measure
- ✓ funnel (a cone made of paper will also work)

### Procedure

1. Fill a small jar about one-quarter full of water.
2. Dip one end of a piece of blue litmus paper into the water.

   **Q1:** What color is the litmus paper when it is wetted by water? What does this indicate?

3. Pour a little club soda into a small cup. Dip one end of a new piece of blue litmus paper into the club soda.

   **Q2:** What color is the litmus paper when it is wetted by club soda? What does this mean?

**Remember...**
- The sample is acidic if blue litmus paper turns red.
- The sample is not acidic if blue litmus paper stays blue.

Page 98

Healthy Air–Pour More Gas

# Pour More Gas

## Leader Guide

**Q3:** The blue litmus paper turns red, meaning that the solution is acidic. The gas from the soft-drink bottle forms an acidic solution when it dissolves in water.

**Q4:** Both litmus papers (from the water containing dissolved gas and from the club soda) turn red.

**Q5:** Some campers may know that carbon dioxide is added to club soda and other carbonated beverages to give these drinks "fizz." This process is called carbonation.

**Q6:** Some campers may remember from the Introducing the Activity discussion that acids typically taste sour. Carbon dioxide gas, when dissolved in the water in mucous and saliva, forms a weak solution of carbonic acid that produces a sour taste.

## Wrap-Up

Ask campers to discuss their answers to the questions. Point out that club soda is made by dissolving carbon dioxide gas into plain water, just like campers did with the jar of water in step 5.

---

### CAMPER

#### Camper Notebook

4. Pour about ⅓ cup (75 mL) vinegar into a soft-drink bottle. Use a funnel to add about 1 teaspoon (5 mL) baking soda into the bottle as shown. Swirl the bottle to make sure the liquid and powder mix well.

5. When the reaction in the bottle slows down, pour some of the gas from the bottle into the jar of water just as you poured the gas into the foil cup containing the lit candle in step 3 of "Pour a Gas." Stop pouring before any of the liquid comes out of the bottle. When you have poured the gas into the jar, seal the jar tightly and shake it vigorously.

6. Now test to see if some of the colorless, odorless gas you made in the soft-drink bottle has dissolved in the water. Open the jar and dip one end of a new piece of blue litmus paper into the water.

**Q3:** What color is the litmus paper when dipped into this solution? What does this mean?

**Q4:** How does the color of the litmus paper from the jar (in step 6) compare to the color of the litmus paper from the cup containing club soda (in step 3)?

**Q5:** What gas is used to make the "fizz" in club soda? (If you don't know, make a guess.)

**Q6:** The gas you made in step 4 by reacting vinegar and baking soda is colorless and odorless, but do you think it is tasteless? Why or why not?
⚠ *Do not taste any of the materials in this experiment.*

# LEADER

## Stirring Indoor Air

### Leader Guide

> ✋ **Suggestions for leading this activity begin on the next page.**

The U.S. Environmental Protection Agency (EPA) ranks indoor air pollution among the top five environmental risks to public health. Studies show that indoor levels of many pollutants can be 2–5 times, and sometimes more than 100 times, higher than outdoor levels. In this activity, campers use boxes and flour to observe the effects of both tightly sealed and well-ventilated buildings on indoor air pollution.

### Some Leader Background Information

Not everyone responds to indoor air pollution the same way. We all know people who are more sensitive to certain substances and cannot tolerate exposure to them, even while the rest of us remain unaffected. How a person responds to indoor air pollutants also depends on age and pre-existing medical conditions. The very young, elderly, and chronically ill are most susceptible to the effects of indoor pollution.

Some pollutants may cause immediate (acute) reactions in sensitive people. Immediate symptoms of exposure to indoor air pollutants can include headaches, tiredness, dizziness, nausea, itchy nose, and scratchy throat. More serious immediate effects are asthma attacks and other breathing disorders. Many immediate health effects are short term and are treatable by removing the source of the contaminant that is causing the symptoms.

The effects of other pollutants are long term (chronic). Long-term health effects occur after years of repeated exposure to a pollutant. These health effects can include serious respiratory diseases and cancer. Indoor air pollutants that cause long-term health effects include environmental tobacco smoke, radon, and asbestos.

Indoor air pollution can come from many sources within a home or building. Pollutants fall into four basic classes: volatile organic compounds (VOCs), biological agents, odorless gases, and particulate matter.

VOCs are a class of chemicals that vaporize easily at room temperature and commonly have strong odors. They are present in many household cleaners, solvents, paints, pesticides, and some cosmetic products. VOCs can also come from new carpeting, synthetic wood products such as particleboard, some types of insulation, and adhesives and other building materials.

Biological agents include molds, bacteria, viruses, dust mites, and animal dander. Molds can be found in damp locations within the home such as bathrooms and basements. Dust mites are tiny spider-like creatures that feed on dandruff and skin; they are present in every home.

Carbon monoxide (CO) and radon are two examples of odorless gases. Carbon monoxide is a product of the incomplete combustion of fuel, and can come from automobile exhaust, wood fires, or improperly adjusted gas- or oil-fired appliances. Radon is a radioactive gas found in some parts of the country that can seep into buildings from the surrounding soil. People exposed to radon for a long period of time have a higher risk of lung cancer.

Particulate matter includes dust and soot from combustion sources such as furnaces and fires. Synthetic carpet that has become old and brittle can also give off large amounts of household dust. Other examples of particulate pollutants include cigarette smoke and asbestos insulation.

Once it is determined that a house has an indoor air quality problem that is affecting its inhabitants, the most effective control method is to attack the problem at its source. For example, depending on the pollutant, you can avoid using pesticides and paints that cause health problems, remove carpeting in bedrooms where occupants have dust mite allergies, or kill mold by removing its nutrient and moisture sources. Bringing in more outdoor air and using air cleaners may also help.

# Stirring Indoor Air

## Leader Guide

## Materials for Camper Procedure—See Left

### Supply Information

- Shoe boxes work well in this activity. Dark interiors are preferred.
- If you cannot find boxes with lids, use aluminum foil for lids. Be sure that the foil overlapping the sides of the box doesn't cover the window openings.
- The clear plastic storage tub should be large enough to hold both cardboard boxes side by side.

### Getting Ready

1. Depending on the abilities of the campers and how much camp time is available, you may want to cut house windows and lid holes prior to the camp session.

### Introducing the Activity

Discuss the health effects and sources of indoor air pollution. (For help with the discussion, see Leader Background Information.) Be sure campers understand that, in this activity:

- The box with plastic wrap windows and tape sealing the lid to the box represents a well-insulated house.
- The box with open windows and an untaped lid represents a house that is not well insulated.
- The flour in this activity represents all indoor air pollutants, including volatile organic compounds (VOCs), biological agents, odorless gases, and particulate matter.
- Air coming from the turkey baster when its bulb is squeezed represents the air circulating inside a house.

Page 223

---

## CAMPER

# Stirring Indoor Air

### Overview

Indoor air pollution is a problem a lot of us never think about, but one that is an increasing concern to many government health agencies. U.S. Environmental Protection Agency (EPA) studies show that the indoor levels of many pollutants can be 2–5 times, and sometimes more than 100 times, higher than outdoor levels. In this activity, you will use flour to represent the spread of air pollutants inside both tightly sealed and well-ventilated model cardboard houses.

### Materials

- ✓ 2 small identical cardboard boxes with lids
- ✓ scissors
- ✓ ruler
- ✓ clear plastic food wrap
- ✓ masking tape
- ✓ paper to make a funnel
- ✓ 6 spoonfuls of flour
- ✓ plastic spoon
- ✓ clear plastic storage tub
- ✓ turkey baster

### Procedure

1. Make two "houses" from cardboard boxes by cutting identically sized windows in the two opposite sides of each box as shown. Cut a hole about the size of a dime in the center of each lid.

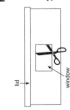

2. On one box, tightly tape plastic food wrap over both windows (so that they are well-sealed all the way around). Tightly seal all seams of the box with tape (including around the edges of the lid). This box represents a tightly sealed house. The other box represents a well-ventilated house.

Page 100

Healthy Air–Stirring Indoor Air

# LEADER

## Stirring Indoor Air

**Leader Guide**

### Procedure Notes and Tips

**Q1:** Flour in the sealed house stays in the sealed house but is stirred around enough to become uniformly distributed. Some of the flour in the house with the open windows escapes to the outside.

**Q2:** The sealed house is more polluted than the open house.

**Q3:** Open the windows or bring in air from the outside to dilute the contaminants. A well-ventilated building can lessen the amount of indoor pollution.

### Wrap-Up

Discuss the results of the activity. Point out that, while the results clearly show that opening windows will reduce the concentration of indoor air pollutants, it does not show that outdoor pollutants can then enter the house through open windows. Outdoor air can contain different pollutants than those found inside, including car exhaust, pollen, and pesticides.

Tell campers that the control of indoor and outdoor pollutants must be a balancing act. People must think about both indoor and outdoor air quality when deciding whether to open windows or not. For example, if a person in the household is allergic to pet dander, opening a window can help by exchanging the air and reducing the amount of indoor dust mite allergens. But if someone in the household is allergic to grass pollen, opening a window can cause problems by bringing more pollen into the house.

Also point out to campers that today's building codes require all new homes to have a special air inlet that brings in fresh air from the outside to mix with the inside air. Heat recovery ventilators (HRVs) can also be used to bring outside air into a building. Both of these methods allow the mixing of indoor air with outside air, thereby reducing the likelihood of indoor pollutants building up within the building.

Page 224

---

# CAMPER

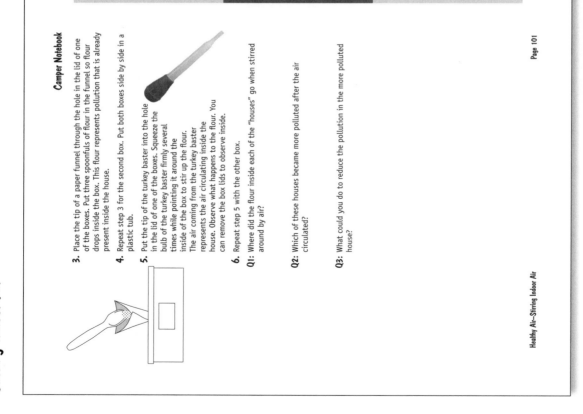

**Camper Notebook**

3. Place the tip of a paper funnel through the hole in the lid of one of the boxes. Put three spoonfuls of flour in the funnel so flour drops inside the box. This flour represents pollution that is already present inside the house.

4. Repeat step 3 for the second box. Put both boxes side by side in a plastic tub.

5. Put the tip of the turkey baster into the hole in the lid of one of the boxes. Squeeze the bulb of the turkey baster firmly several times while pointing it around the inside of the box to stir up the flour. The air coming from the turkey baster represents the air circulating inside the house. Observe what happens to the flour. You can remove the box lids to observe inside.

6. Repeat step 5 with the other box.

**Q1:** Where did the flour inside each of the "houses" go when stirred around by air?

**Q2:** Which of these houses became more polluted after the air circulated?

**Q3:** What could you do to reduce the pollution in the more polluted house?

Page 101

Healthy Air—Stirring Indoor Air

# Trapping Particulates

## Leader Guide

👉 Suggestions for leading this activity begin on the next page.

Volatile organic compounds (VOCs), inorganic gases, biological contaminants, and particulates are the main types of contributors to poor indoor air quality. Of these air contaminants, VOCs (some of which have detectable odors) and inorganic gases (some of which are odorless) are small molecules which are usually too small to trap in filters. Biological contaminants (see examples in table at right) are larger and may be trapped by some filters. In this activity, campers explore vacuum bags to compare their ability to trap finely divided powders. Campers will realize that some sources of poor indoor air quality can be improved by cleaning and filtering with appropriate materials.

## Some Leader Background Information

Some bedding covers, furnace filters, vacuum cleaner bags, and air purifier systems with filters claim to trap nearly all dust and allergens. The holes in these products and filter systems are generally small enough that many biological indoor air contaminants are trapped within. The table at right lists some approximate sizes of biological sources contributing to poor indoor air quality.

In this activity, you may discover that not all home filtering devices are equal. Typically, plain vacuum cleaner bags that don't make specific filtration claims will eventually become loaded with contaminants, allowing these contaminants to go through the sides of the bags and back into the room. In comparison, vacuum cleaner bags advertising such claims as "micro filtration," "allergen filtration," and "HEPA filtration" will trap fine particles much more efficiently. In our testing, these bags did not allow cornstarch, talcum powder, cinnamon, or fine charcoal to pass through the bags' sides even after 300 taps.

For a filter to be called a High-Efficiency Particulate Air (HEPA) filter, it must be 99.97% efficient at filtering particles having a particle size of 0.3 µm. Most biological indoor air pollutants are likely to be trapped by a HEPA-rated bag or filter and removed from the environment.

Keep in mind that, in order for any filter to work, the contaminants have to get to the filter. Filtering every bit of air in a room or vacuuming every room surface is not very feasible. Because of this, filters can help reduce some particulate matter and biological contaminants but will probably not eliminate them entirely.

Be aware that this activity does not take into account the amount of air going through the filter. For example, plain, non-allergen vacuum bags may allow more air to pull through the vacuum cleaner due to the larger pore size of the bags. It is usually assumed that filters with small pore sizes still allow for optimal performance of the cleaning device, but whether or not this is true is not established in this activity.

| Biological Contaminant | Approximate Average Size |
|---|---|
| bacteria—*E. coli* | 2 µm |
| bacteria—*Staphylococcus* | 0.5 µm |
| cat dander | 5–6 µm |
| dog dander | larger than 5–6 µm |
| dust—human dander | 10 µm |
| dust mite allergens—fecal matter | 6–30 µm |
| dust mites | 250–300 µm |
| mold spores | 15 µm |
| pollen—ragweed | 20–120 µm |
| pollen—typical | 5–200 µm |
| virus—vaccinia | 0.2 µm |

**FYI...**
- µm is the symbol for micrometer or micron, which is a unit of length that is one millionth of a meter.
- The diameter of a human hair is about 70 µm.

# LEADER

## Trapping Particulates — Leader Guide

### Materials for Camper Procedure—See Left

### Supply Information

- When selecting the disposable vacuum cleaner bags to use in this activity, read the packages of generic brands, store brands, and name brands. Some filters will be advertised as having "micro filtration," "allergen filtration," "99.7% filtration," and "HEPA filtration." Be sure campers test at least one filter that makes these claims and at least one filter that does not.

### Procedure Notes and Tips

- Be sure to save all vacuum cleaner bag packages and sale slips so campers can refer to them during the session.
- A long rectangular bag (for an upright vacuum cleaner) with an attachment opening near the center of the bag can be cut into two pieces so both halves can be used by campers.
- If you'd like, different campers can try different powders and then campers can compare their results.
- The powder that comes through the bags is very fine. When using black paper, it may be easier to notice the powder that has gone through the bag by wiping the paper's surface with a finger and then looking on the finger for fine powder residue.

---

# CAMPER

## Trapping Particulates

### Overview

Particulate matter (such as dust and soot) and biological contaminants (such as mold, pollen, animal dander, and bacteria) get into our homes and can cause poor indoor air quality. Can we reduce or eliminate these contaminants by ordinary cleaning methods like vacuuming?

### Materials

- ✓ two or more different types of disposable vacuum cleaner bags
- ✓ scissors
- ✓ packages from the vacuum cleaner bags used in this activity
- ✓ sales slip listing the cost of each package of vacuum cleaner bags
- ✓ fine powder (such as cornstarch, talcum powder, cinnamon, or activated charcoal)
- ✓ teaspoon or other volume measure
- ✓ tape
- ✓ paper having a color that contrasts with the fine powder
  *For example, use black paper if testing with cornstarch or talcum powder and white paper if testing with activated charcoal or cinnamon.*

### Procedure

1. Cut the attachment opening off each vacuum cleaner bag so that you are left with bag-like filters having one opened end. (If the vacuum bag is long and the attachment opening is near the middle, both the top and bottom of the bag can be used. Share the other half with another person or group.)

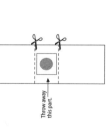

Throw away this part.

Throw away this part.

Healthy Air—Trapping Particulates

# Trapping Particulates

## Leader Guide

**Q1:** Answers will vary depending on the brands tested. Some campers may conclude that the bags designated for trapping allergens do a better job of trapping the fine powder than the plain bags simply labeled "filter bags."

**Q2:** Answers will vary. If the goal of the filter is to trap all possible dirt and allergens, the allergen-filtering bags are better, even though they cost about twice what a regular bag costs.

## Wrap-Up

Ask campers to share their results. Discuss the approximate sizes of common biological contaminants found in the home. Relate these sizes with pore sizes of vacuum bags and other household filters. (For help with the discussion, see Leader Background Information.)

---

### CAMPER

#### Camper Notebook

2. In the data table at the end of this activity, record the type of powder and the brands of bags you will test. List any performance claims printed on the package (such as "stress tested," "allergen reduction," or "HEPA filter") and the cost per bag of each brand. Divide the cost of the package by the number of bags in the package to get the cost per bag.)

3. Look closely at the construction of each brand of bag, including its thickness and the material it's made of. Describe each bag's construction in the data table.

4. Add 1 teaspoon (5 mL) fine powder to each bag. Fold the tops of the bags down twice to make the seals powder tight. Tape the folds closed.

5. Count the taps as you tap each bag repeatedly on the table over a piece of paper. As soon as the powder works its way through the bag and onto the paper, stop tapping and record the number of taps in the data table. If no powder is observed after 150 taps, record "trapped after 150."

**Q1:** Which bag did the best job of trapping very fine powder?

**Q2:** Considering the cost and filtering abilities of the different bags, which bag would you buy and why?

Powder Tested:

| Brand of Vacuum Bag | Performance Claim | Cost per Bag | Bag Construction | Number of Taps |
|---|---|---|---|---|
| | | | | |
| | | | | |
| | | | | |
| | | | | |
| | | | | |

## LEADER

# Cartesian Diver

## Leader Guide

> 👉 **Suggestions for leading this activity begin on the next page.**

In this activity, campers make a toy called a Cartesian diver to explore the behavior of gases. Specifically, they investigate how the compressibility of air inside a Cartesian diver affects its movements in water.

## Some Leader Background Information

In the mid 1600s, an English scientist named Robert Boyle experimented with gases and noticed their behavior under changing circumstances. Boyle's Law states that, at a constant temperature, the pressure and volume of a given amount of gas are closely related. If one goes up, the other must go down. Mathematically, $pV = $ constant, where $p$ is the pressure and $V$ is the volume. This means if the pressure is doubled, the volume will be halved (if the temperature remains the same).

Objects either sink or float in water because of their densities. Objects more dense than water will sink, and objects less dense than water will float. The diver in this activity is a multipart system. By itself, the hex nut is more dense than water and therefore sinks. When the hex nut is attached to the pipet containing air, the resulting system is less dense than water and so it floats.

In this activity, Boyle's Law accurately describes the Cartesian diver's behavior. The large amount of water in the bottle and the small size of the diver's bulb mean that the temperature of the entire system remains essentially constant. Squeezing the sides of the bottle increases the pressure of the gas inside the bottle and inside the diver. The gas compresses, its volume is reduced (as per Boyle's Law), and more water flows into the bulb of the diver. The mass of the diver increases with the added water, its density increases, and the diver sinks.

When the pressure on the outside of the bottle is released, the pressure of the gas in the diver is reduced, the volume of the gas in the diver increases (as per Boyle's Law), and water is forced out of the diver. Since the mass of the diver is now less (since there is less water in the diver), the diver's density decreases and the diver rises to the surface of the water.

Putting one piece of glitter into the 1-L bottle of water helps campers visualize a concentration of 1 part per million (ppm). One piece of glitter (which represents a contaminant) has a mass of about 0.001 g. At 4° C, 1.000 L water has a mass of 1000 g. Therefore, the 0.001 g glitter in 1000 g water is equal to 1 mg in 1,000,000 mg or 1 ppm.

# Cartesian Diver

## Leader Guide

### Materials for Camper Procedure—See Left

### Introducing the Activity

You may want to demonstrate how to make the diver. (See steps 1 and 2 in the Camper Notebook.) Also, you may want to show campers how to adjust the amount of water in the diver's pipet bulb (as described in step 3) so that it floats in the cup of water.

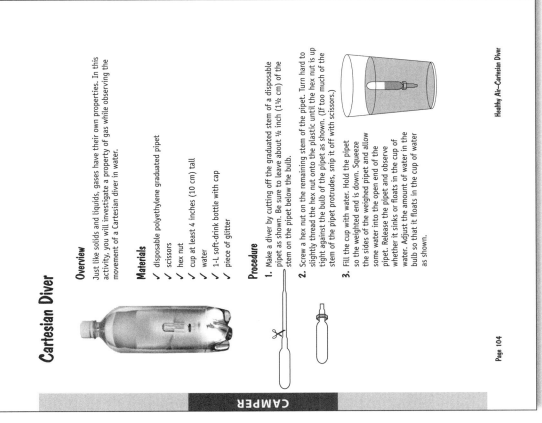

## Cartesian Diver

### Overview

Just like solids and liquids, gases have their own properties. In this activity, you will investigate a property of gas while observing the movement of a Cartesian diver in water.

### Materials

- disposable polyethylene graduated pipet
- scissors
- hex nut
- cup at least 4 inches (10 cm) tall
- water
- 1-L soft-drink bottle with cap
- piece of glitter

### Procedure

1. Make a diver by cutting off the graduated stem of a disposable pipet as shown. Be sure to leave about ½ inch (1½ cm) of the stem on the pipet below the bulb.

2. Screw a hex nut on the remaining stem of the pipet. Turn hard to slightly thread the hex nut onto the plastic until the hex nut is up tight against the bulb of the pipet as shown. (If too much of the stem of the pipet protrudes, snip it off with scissors.)

3. Fill the cup with water. Hold the pipet so the weighted end is down. Squeeze the sides of the weighted pipet and allow some water into the open end of the pipet. Release the pipet and observe whether it sinks or floats in the cup of water. Adjust the amount of water in the bulb so that it floats in the cup of water as shown.

Healthy Air—Cartesian Diver

# LEADER

## Cartesian Diver

## Leader Guide

### Procedure Notes and Tips

**Q1:** When the bottle is squeezed, the diver sinks in the water. As the bottle is squeezed harder, the diver sinks faster. As the bottle is released, the diver floats to the top again.

**Q2:** The volume of gas (air) in the diver decreases as the bottle is squeezed, so more water flows into the diver.

**Q3:** Answers may vary. The diver is open at the bottom and water flows into the diver when its gas is compressed (when the bottle is squeezed). If the glitter particle is floating in the water near the end of the diver, it is possible for the glitter to end up inside the diver as gas is compressed in the diver and water flows in. If the glitter is sitting at the bottom of the bottle, the diver won't be able to pick up the glitter. With practice, campers may be able to get the sinking diver to land on the glitter, but releasing the bottle to allow the diver to float results in water flowing out of the diver. This keeps the glitter from being picked up by the diver.

### Wrap-Up

Ask campers to share their answers to the questions. Discuss how the movement of the Cartesian diver within the bottle relates to Boyle's Law. (For help with the discussion, see Leader Background Information.)

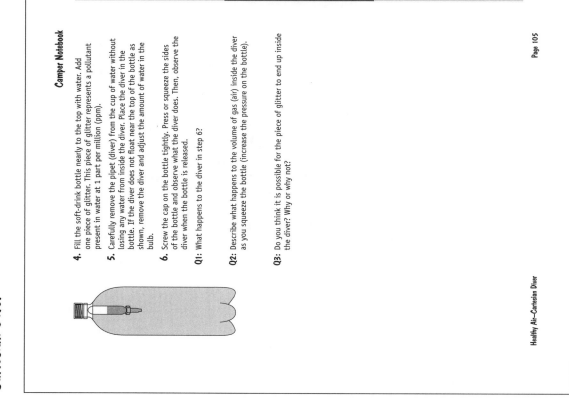

## CAMPER

### Camper Notebook

4. Fill the soft-drink bottle nearly to the top with water. Add one piece of glitter. This piece of glitter represents a pollutant present in water at 1 part per million (ppm).

5. Carefully remove the pipet (diver) from the cup of water without losing any water from inside the diver. Place the diver in the bottle. If the diver does not float near the top of the bottle as shown, remove the diver and adjust the amount of water in the bulb.

6. Screw the cap on the bottle tightly. Press or squeeze the sides of the bottle and observe what the diver does. Then, observe the diver when the bottle is released.

**Q1:** What happens to the diver in step 6?

**Q2:** Describe what happens to the volume of gas (air) inside the diver as you squeeze the bottle (increase the pressure on the bottle).

**Q3:** Do you think it is possible for the piece of glitter to end up inside the diver? Why or why not?

# Leader Guide

# Growing Mold

✋ Permission is granted for you to copy and distribute this take-home activity for your event. (The master is provided in the Camper Notebook.)

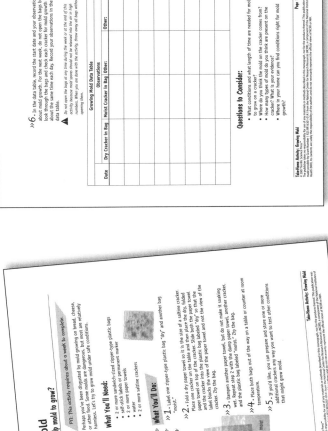

Healthy Air—Growing Mold

Page 231

LEADER

# TOPiC 4

## HEALTHY SKIN

### List of Activities

- ✓ Hydrophobic Art .................................................. 234
  *Explore hydrophobic paint to learn how skin serves as a protective layer for the outside of the body.*

- ✓ Visible Light Challenge ......................................... 237
  *Investigate the relative energy of visible light using different colors of light.*

- ✓ How Sensitive Is Your Skin? .................................... 239
  *Measure the sensitivity of different parts of the hand.*

- ✓ Make and Test Lip Balm ........................................ 242
  *Create and test several lip balm products to assess their sunscreen and waterproof properties.*

- ✓ Cover Up, Screen, or Block? ................................... 247
  *Discover the UV protection of certain clothing, sun products, and sunglasses.*

- ✓ Sunning Straws .................................................. 252
  *Use benzophenone and a UV light reaction to quantitatively measure the amount of UV protection that different sun protection products provide.*

- ✓ Suntan in a Bottle .............................................. 255
  *Investigate how a sunless tanning product reacts with different natural and synthetic fabrics.*

- ✓ A Look at Bleaching ............................................ 259
  *Observe the effects of bleach on fabric to mimic what skin bleaching products might do to skin.*

- ✓ Take-Home Activity: UV Detective Challenge ............... 263
  *Make a UV detection bracelet, then place the bracelet in different outdoor locations to test levels of UV radiation.*

# LEADER

## Leader Guide

# Hydrophobic Art

> ✋ **Suggestions for leading this activity begin on the next page.**

Skin serves as a protective layer for the outside of the body. In this role, skin may allow some liquids out of the body (such as sweat), but it usually holds other body fluids inside. Skin is mostly hydrophobic, meaning that it is "water hating" (it lacks affinity for water). Notice that rain drops run off the outside of skin. In this activity, campers explore hydrophobic paint while decorating a pencil.

## Some Leader Background Information

When a small amount of oil-based enamel paint is placed on the surface of water, the paint tries to spread in a thin layer over the entire surface. This behavior occurs because the paint is hydrophobic. ("Hydro" is Greek for "water" and "phobic" is Greek for "fearing.") Hydrophobic substances do not mix with water. Conversely, a substance that is attracted to water is hydrophilic. ("Philic" is Greek for "loving.")

The hydrophobic oil-based paint used in this activity repels water due to the oil's chemical composition (mostly the elements carbon and hydrogen). Carbon and hydrogen atoms in the oil molecules* equally share electrons in bonds. In contrast, the chemical elements in water (oxygen and hydrogen atoms) share electrons unequally. Oxygen has a greater portion of electron density in oxygen-hydrogen bonds. This unequal electron density distribution in a bond, as well as the geometric arrangement of the oxygen and hydrogen atoms in space, causes the water molecule to behave like a magnet with polarized ends. These polarized ends cause water molecules

to stick together. Water is said to be polar. Other chemical substances are similarly polar, hydrophilic, and readily dissolve in water. Chemical substances that are not polarized (such as oil-based paints) are nonpolar, repel water, and are hydrophobic.

Some enamel paint is oil-based paint. Because oils consist mostly of carbon and hydrogen atoms, with almost equally shared electrons, the bonds are not very polarized. Oils are nonpolar and do not dissolve in water (which is polar). The polar-polar attraction of one water molecule toward another water molecule is much stronger than the nonpolar-nonpolar attraction of one oil molecule toward another oil molecule. Also far weaker than the polar-polar attraction of water molecules toward other water molecules is the smaller attraction of the polar water toward nonpolar oil. This weaker polar-nonpolar interaction causes the oil to spread into a thin layer (possibly a monolayer only one molecule thick) upon the surface of the water.

---

*The term "molecule" is used for both oil and water particles in this discussion since both particles are in fact molecules. If you use this background material in a discussion with campers, the term "particles" may be used without distinguishing between molecular and ionic compounds.

**Healthy Skin—Hydrophobic Art**

# Hydrophobic Art

## Leader Guide

### Materials for Camper Procedure—See Left

### Materials for Getting Ready

✓ clear plastic 1- or 2-L soft-drink bottle (one per camper or group)
✓ scissors

### Getting Ready

1. Cut the tops off the soft-drink bottles as shown.

### Introducing the Activity

Discuss safety issues related to using enamel paint and oil-paint solvent. (For help with the discussion, see Materials in the Camper Notebook and see the labels on the packages.)

### Procedure Notes and Tips

- Planning ahead with extra staff members and organization may be helpful in reducing paint stains.
- Oil-based paint can stain clothing and furniture. Extra care should be taken to prevent paint stains. Tell participants to wear old clothing, or provide smocks or aprons. Have turpentine or another appropriate oil-paint solvent available to clean up paint spills during the activity.
- Have campers work in a well-ventilated area and observe the other safety precautions.
- Work over newspaper at all times.
- If more than one camper needs to share the same cut-off soft-drink bottle, be sure campers add additional paint to the water as needed before dipping additional pencils.

Page 235

---

## Hydrophobic Art

### Overview

Skin serves as a protective layer for the outside of the body. Skin allows some liquids out of the body (such as sweat), but it usually holds other body fluids inside. Skin is mostly hydrophobic, meaning that it is "water hating" (it lacks affinity for water). Notice that rain drops run off the outside of our skin. Explore another hydrophobic material in the following activity as you decorate a pencil to take home.

### Materials

✓ newspaper and paper towels
✓ clear plastic 1- or 2-L soft-drink bottle (prepared by leader)
✓ water
✓ toothpicks
✓ 2 different colors of Testors brand oil-based enamel model paint
⚠ *Work in a well-ventilated area. Enamel paint is flammable, so avoid sparks and open flames. Avoid getting paint on the skin. If you do, use an oil-paint solvent to remove the paint and then wash the skin with plenty of soap and water.*
✓ masking tape
✓ wooden pencil or wood cut-out
✓ spring-type clothespin
✓ oil-paint solvent (such as turpentine) for emergency clean-ups
⚠ *Handle the solvent as directed on its package label.*

### Procedure

1. Lay newspaper over the work area. Fill the cut-off soft-drink bottle with water as shown.
2. Using the precautions listed above for handling enamel paint, take a clean toothpick and drop about 4–5 drops of one color of enamel paint on the water. Notice what happens to the paint.

Q1: Does the paint sink or float? Does it mix with the water?

Page 110

**CAMPER**

Healthy Skin—Hydrophobic Art

# LEADER

## Hydrophobic Art

### Leader Guide

- Campers can also dip unfinished wood cut-outs, which are available at craft and hobby stores. Clip a clothespin to the very edge of the wood cut-out. Have campers hold the clothespin while dipping the cut-out into the marbling mixture.

- When the activity is done, skim paint from the water's surface using a piece of newspaper or a paper towel. Dispose of newspapers and paper towels in the trash. Dispose of the water down the drain.

**Q1:** The enamel paint floats on the water and does not mix with the water.

**Q2:** Some of the paint sticks to the toothpick as the toothpick is submerged in the water.

### Wrap-Up

Relate the hydrophobic characteristics of skin to the same characteristics of enamel paint. Based on the age and ability of the campers, you may want to discuss the difference between polar and nonpolar substances. (For help with the discussion, see Leader Background Information.)

---

## CAMPER

### Hydrophobic Art

**Camper Notebook**

3. Add 4–5 drops of a second color of enamel paint using a clean toothpick. Then, use this toothpick to slowly swirl the paint into a marbled pattern.

4. Dip one end of another clean toothpick about halfway into the marbling mixture. Look through the side of the bottle when you do this. Then remove the toothpick and notice the effect on the toothpick.

**Q2:** What do you observe as the toothpick is dipped into the water?

5. Write your name on a piece of tape and attach the tape to the eraser end of a pencil as shown.

6. Holding the pencil by the eraser, dip the pencil into the marbling mixture while slowly twirling the pencil as shown. (Twirling the pencil while dipping prevents paint from clumping together in one spot and produces a more even marbling effect.)

7. Remove the pencil from the water after it is dipped down to the label as shown. Clip a clothespin onto the taped end of the pencil. Prop the pencil over newspaper and paper towels by using the pencil's lead end and the clothespin's two open feet as shown. Allow the pencil to completely dry.

8. If another person needs to dip his or her pencil into the same container, repeat steps 2 and 3 to add more paint to the water as needed.

*Healthy Skin—Hydrophobic Art*

# Visible Light Challenge

👉 **Suggestions for leading this activity begin on the next page.**

Ultraviolet (UV) radiation is a dangerous, high-energy form of light that is invisible to the human eye. UV is just one type of energy that scientists call electromagnetic radiation. Depending on the wavelength (how closely spaced or stretched out the crests of the waves are), we see electromagnetic radiation as visible light, feel its presence as heat, receive it through a radio as sound, or use it to cook our food in a microwave oven. (See figure at right.) In this activity, campers learn about the relative energy of visible light, the part of electromagnetic radiation that can be seen.

## Some Leader Background Information

Visible light is made up of the familiar spectrum of colors from violet to red (colors of the rainbow). UV is the more energetic form of light just past violet in the electromagnetic spectrum. Humans cannot sense UV radiation until it has done its damage and we feel its effect as sunburn.

Scientists divide UV into three types: UVA, UVB, and UVC. UVA is the least energetic and probably the least dangerous form of UV radiation. Most UVA passes through the atmosphere unfiltered and makes up about 95% of UV radiation reaching the earth's surface. But "least dangerous" does not mean safe. Some researchers suspect that UVA may be responsible for malignant melanoma, a deadly form of skin cancer. UVB, the next most energetic form of UV radiation, is partially

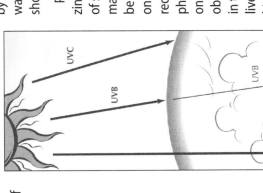

absorbed by the ozone layer. Generally only the longest wavelengths of UVB reach the earth's surface. UVB is responsible for sunburns and most tans. The atmosphere completely blocks out UVC (the most energetic type of UV radiation), so UVC does not reach the earth's surface.

Since we can not see UV radiation, campers use visible light given off by different-colored light emitting diodes (LEDs) to learn how different wavelengths of electromagnetic radiation have different energies. (The shorter the wavelength, the higher the energy.)

Phosphorescent yellow green glow-in-the dark vinyl sheets often contain zinc sulfide doped with copper. When a set of red, green, and blue LEDs of similar size and shape are used to excite the same phosphorescent material, the relative energies of the different wavelengths of the light can be determined. In a darkened room, when the red LED is placed directly on the phosphorescent sheet, no phosphorescent glow is observed. The red light emitted by the LED does not have sufficient energy to excite the phosphorescent sheet. When the green LED is placed in the same manner on the same phosphorescent sheet, a faint glow-in-the-dark effect is observed. Finally, placing the blue LED on the same phosphorescent sheet in the same manner as the red and green LEDs causes a bright and long-lived phosphorescent trail. These results demonstrate that the blue light is highest in energy of the three colors.

## LEADER

# Visible Light Challenge    Leader Guide

## Materials for Camper Procedure—See Left

### Supply Information

- Different-colored LED flashlights can be purchased in variety and specialty stores that sell items such as electronic supplies, novelty toys, automotive supplies, and fishing gear.
- Phosphorescent "glow-in-the-dark" vinyl sheets can be found in science teacher supply stores (especially on the Internet). As an alternative, glow-in-the-dark vinyl shapes (such as stars and planets) can be used. A less effective method is to paint something with glow-in-the-dark paint and let dry before using it in the activity.

### Procedure Notes and Tips

**Q1:** When the same phosphorescent sheet is tested with the red, green, and blue LEDs, the phosphorescent trail left by the blue LED is the strongest and lasts the longest. Therefore, the blue light must have the highest energy. The red light must have the lowest energy because no trail is observed.

### Wrap-Up

Review camper results and the relative energy of the LEDs that were tested. Discuss electromagnetic radiation, visible light, and UV radiation. (For help with the discussion, see Leader Background Information.)

---

## CAMPER

# Visible Light Challenge

### Overview

UV radiation is high in energy and therefore damaging to skin. Although we cannot see UV radiation, this activity uses visible light of different colors to introduce the idea that different wavelengths of light have different levels of energy. Which of the colors has the highest energy?

### Materials

- ✓ darkened room
- ✓ red, green, and blue light emitting diode (LED) mini flashlights
- ✓ phosphorescent (glow-in-the-dark) vinyl sheet
- ✓ timer such as stopwatch or watch with second hand

### Procedure

1. In a darkened room, turn on either the green or blue LED mini flashlight. Hold the flashlight against the green side of the phosphorescent sheet so that the light is touching the sheet. Start timing and move the flashlight around on the phosphorescent sheet for 10 seconds. With the timer still ticking, turn off the flashlight and time how long the trail is visible on the phosphorescent sheet.
2. Record your results in the data table below.
3. Repeat steps 1 and 2 with the same phosphorescent sheet and the other two LED mini flashlights (one at a time).

**Q1:** Based on your results, which of the emitted lights do you think has the highest energy and which has the lowest energy? Why?

| LED Color | Phosphorescent Trail? (Yes or No) | How Long the Trail Lasted |
|---|---|---|
|  |  |  |
|  |  |  |
|  |  |  |

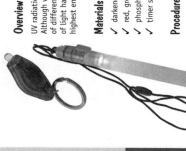

**Leader Guide**

# How Sensitive Is Your Skin?

> **Suggestions for leading this activity begin on the next page.**

Some people who don't use sunscreens or sunblocks may think that they will "feel" when they have had enough sun. The problem with this theory is that sunburn has already happened by the time they feel it. Skin contains neurons that respond to the sense of touch. In this investigation, you will measure the sensitivity of different parts of the hand.

## Some Leader Background Information

In the late 1800s, German physiologist Max von Frey used hairs of different diameters to map skin sensitivity. He mounted human hairs and other bristles on handles and used them to identify specific pain spots on the back of the hand. This technique, and variations of it, are still being used in neurology today.

In this activity, campers make a set of von Frey hairs out of different thicknesses of fishing line. They investigate to determine the tactile detection thresholds of different parts of the hand. Depending on the hand locations that are chosen and on the person being tested, campers should discover that some parts of the hand are more sensitive to touch than other parts. Generally, the center of the back of the hand is less sensitive than the palm of the hand. The fingertips tend to be more sensitive than other parts of the hand.

# LEADER

## How Sensitive Is Your Skin?
Leader Guide

### Materials for Camper Procedure—See Left

### Introducing the Activity
Introduce the scientific research of Max von Frey. (For help with the discussion, see Leader Background Information.) Describe what campers need to do to make their own von Frey hairs.

### Procedure Notes and Tips

- Make sure campers test with a wide range of fishing line thicknesses.
- Make sure campers touch the line to one point on the skin and do not move the line across the skin. This test should analyze the feeling of pressure on the skin caused by the pressure of the line.
- Warn campers that touching the line to a part of the skin containing fine hairs may make the skin appear more sensitive. Movement of the hairs on the skin amplifies the touch of the line.
- The University of Washington website offers many touch experiments. To try some, go to www.washington.edu and search "Chudler touch experiments."

---

# CAMPER

## How Sensitive Is Your Skin?

### Overview
Some people who don't use sunscreens or sunblocks may think that they will "feel" when they have had enough sun. The problem with this theory is that sunburn has already happened by the time they feel it. Skin contains neurons that respond to the sense of touch. In this investigation, you will measure the sensitivity of different parts of the hand with a set of testing devices called von Frey hairs.

### Materials
- ✓ 5 different thicknesses of fishing line having a wide range of diameters
- ✏️ *Transparent sewing thread may be used for a thin diameter.*
- ✓ scissors
- ✓ tape
- ✓ craft sticks
- ✓ (optional) blindfold

### Procedure

1. Make your own set of von Frey hairs with different thicknesses of fishing line. For each thickness, cut a length of fishing line about 5 cm (2 inches) long. Tape the end of the fishing line onto a craft stick at a 90° angle as shown. Label the craft stick with the diameter of the fishing line.

2. Working in pairs or small groups, decide on up to five parts of the hand that you want to test. Circle and number these locations in the hand diagrams at the end of this activity. Record your fishing line thicknesses in the data table at the end of this activity by listing the thinnest size to the thickest size.

3. Decide on the person whose hand will be tested. Blindfold that person or have him or her look away. Touch the thinnest fishing line to the first hand location until the fishing line bends. Do not move the fishing line along the skin. Ask the person if he or she feels anything and record the answer in the data table. Repeat the procedure with the same fishing line for all hand locations.

# How Sensitive Is Your Skin? — Leader Guide

**Q1:** Generally, different parts of the hand have different tactile detection thresholds. The center of the back of the hand is less sensitive than the outside of the back of the hand. The finger tips are designed for touching, so they often are most sensitive. Different individuals may have different tactile detection thresholds.

**Q2:** Campers should report results that are consistent with their data. Often the finger tips have the lowest tactile detection threshold, and are the most sensitive.

---

## Camper Notebook

4. Repeat step 3 for all fishing line thicknesses. If time permits and if you want, test the hands of other group members and record their answers in the data table using a method that distinguishes one set of answers from another (such as different ink color, cursive writing, all capital letters, or something else).

5. Tactile detection threshold is the smallest amount of touch necessary for a person to feel. For example, imagine touching four different thicknesses of fishing lines (0.30 mm, 0.41 mm, 0.61 mm, and 0.89 mm) on the tip of an index finger. If the person only felt the 0.61 mm and 0.89 mm thicknesses, the tactile detection threshold would be 0.61 mm (the smallest diameter of line felt). In your data table, circle the tactile detection threshold for each hand location you tested (and for each person's hand you tested).

**Q1:** Look at the data table. If the detection threshold is different for different parts of the hand, describe the differences.

**Q2:** Where is the most sensitive part of the hand that you tested?

**FYI...** Circle and number hand locations on these hand diagrams.

palm · back

| Write fishing line thicknesses across this row (thinnest to thickest). | Was the touch felt with the following fishing line thickness? | | | |
|---|---|---|---|---|
| location 1 | | | | |
| location 2 | | | | |
| location 3 | | | | |
| location 4 | | | | |
| location 5 | | | | |

## LEADER

*Leader Guide*

# Make and Test Lip Balm

> 👉 Suggestions for leading this activity begin on the next page.

Commercial lip balm is made to protect lips from wind and other drying conditions. Some lip balms also contain sunscreen. In this activity, campers make lip balm from a few natural oily and waxy ingredients. Oil of cinnamon is used for favor. Then, campers create several test products using different concentrations of the same ingredients and assess these test products for sunscreen and waterproof properties.

## Some Leader Background Information

A mixture of natural oily and waxy materials can be combined to create a spreadable mixture that is similar to commercial lip balms. Lanolin, or hydrous lanolin, is a waxy ointment base isolated from wool. Coconut oil is an oil that is hydrophobic (water-hating). Beeswax is a waxy solid that is also hydrophobic. Oil of cinnamon is the essential oil of cinnamon bark. A mixture of these oils and waxes has a melting point higher than that of the oils but lower than that of the waxes. These hydrophobic materials applied to the surface of the skin help the skin retain moisture when exposed to wind and other drying conditions. Saturated oils and waxes (such as coconut oil and beeswax) are less prone to oxidation and may increase the stability of the product.

In this activity, campers observe that increasing amounts of oil of cinnamon in the test products result in lighter UV detection beads and therefore greater screening of UV light. However, campers also observe that the sunscreen agent (oil of cinnamon) migrates from the balm into the cup of water, indicating that the sunscreen is not waterproof. This observation should lead campers to conclude that none of these test products would serve as a durable sunscreen.

# Make and Test Lip Balm

## Materials for Camper Procedure—See Left

### Supply Information

- 100% lanolin is available from health food stores.
- 100% coconut oil is available from health food and grocery stores.
- Beeswax is available where candle-making supplies are sold.
- Food-grade oil of cinnamon is available where baking supplies are sold.

### Introducing the Activity

Discuss the oily and waxy ingredients that campers will use to make their lip balm. (For help with the discussion, see Leader Background Information.) Point out that oil of cinnamon will be used as a flavor. One of the active sunscreen ingredients approved for use in the United States is octyl methoxycinnamate. Ask the campers whether they think oil of cinnamon would be effective in place of the octyl methoxycinnamate in lip balm. Tell campers that they will make various test products to address this question. Remind campers that an effective sunscreen formulation must screen out the sun's UV radiation and must be waterproof.

⚠️ *Warn campers that the test products they make after step 4 will contain high concentrations of oil of cinnamon that are not approved for use on lips and skin. Oil of cinnamon may be irritating at high levels. In the real world, only if oil of cinnamon is better than existing sunscreen products at UV screening and waterproof ability (as well as other requirements) would someone consider testing the products on animals and humans. Since very strict rules govern animal and human testing, campers should not apply test products to their skin and lips.*

---

# Make and Test Lip Balm

## Overview

In this activity, you will make lip balm using oil of cinnamon as a flavor. You will also create test products having varying concentrations of oil of cinnamon and then test these products for sunscreen and waterproof abilities.

## Materials

- ✓ self-stick labels
- ✓ permanent marker
- ✓ 5 or 6 mini foil muffin cups
- ✓ lanolin
- 🔘 balance capable of measuring 0.1 g or set of measuring spoons
  *Measuring ingredients is more accurately done by mass with a balance rather than by volume with measuring spoons.*
- ✓ coconut oil
- ✓ beeswax
- ✓ electric food-warming tray
  ⚠️ *Use the warming tray carefully and with adult supervision.*
- ✓ toothpicks
- ✓ oil of cinnamon
  ⚠️ *Large amounts of oil of cinnamon can be irritating to skin.*
- ✓ dropper or disposable pipet
- ✓ 12 paper twist ties
- ✓ 12 UV detection beads of the same color
- ✓ waxed paper
- ✓ cup of water
- ✓ (optional) knife to break up the beeswax
- ✓ (optional) paper such as paper muffin cup
- ✓ (optional) small plastic knife, spatula, or bowl scraper
- ✓ (optional) towel to wrap around electric food-warming tray
- ✓ (optional) UV lamp
  ⚠️ *Looking directly at the UV light can cause eye damage.*

# LEADER

## Make and Test Lip Balm

### Leader Guide

### Procedure Notes and Tips

- The use of taring, top-loading digital balances can make this activity easier, since two to three weighings are required per product. If offering these balances, explain to campers how to tare the balance.

- Campers using a balance to measure the lanolin and coconut oil in step 1 can place the foil muffin cup directly on the balance. Each substance can be added one at a time to a different spot in the muffin cup so that excess amounts can be removed without contamination.

- If ingredients are measured by volume, designate a measuring spoon of the appropriate volume for each ingredient. Each camper can use the same measuring spoon for that ingredient. Since many ingredients are sticky, a small plastic knife, spatula, or bowl scraper may be useful as a measuring spoon scraper.

- If using a balance to measure the beeswax, place paper (such as a paper muffin cup) on the balance and then add the beeswax to measure its mass.

- If using a teaspoon to measure the beeswax, either melt the beeswax in a clean mini foil muffin cup before measuring or tightly pack shavings (or pellets) into a teaspoon.

- The use of an electric food-warming tray will allow many muffin cups to be warmed together rather quickly. The hot tray should be attended at all times. The small volumes and short heating times mean campers will be able to lift the muffin cups by the edges after the mixtures are melted. For younger campers or for extra safety, towels can be wrapped around the edges of the hot electric food-warming tray.

- If the beads are not completely covered after step 10, the beads can be dipped again to make a second coat. Even if only some beads need second coats, dip all of the beads again for consistency. Do not dip any beads a third time.

---

# CAMPER

## Make and Test Lip Balm

### Camper Notebook

### Procedure

1. Label a mini foil muffin cup with your name and "LB" to stand for "lip balm." Add 1.0 g (¼ teaspoon) lanolin, 5.0 g (1¼ teaspoon) coconut oil, and 4.0 g (1 teaspoon) beeswax according to your leader's instructions.

2. Heat the muffin cup on an electric food-warming tray until all ingredients are melted. Stir occasionally with a toothpick.

3. Remove the muffin cup from the warming tray and add 2 drops oil of cinnamon. Stir. Allow the mixture to cool to room temperature and solidify.

4. If you are not allergic to any of the ingredients used in steps 1 and 3, you may test the cooled lip balm on your fingers first, and then apply to your lips. You may take your lip balm home.

**Q1:** How would you rate your lip balm compared to other lip balms you have used?

5. Now you will make four different test products. Begin by labeling four mini foil muffin cups with "10," "20," "30," and "40." (The numbers represent the different concentrations of oil of cinnamon that you will test.) Add 1.0 g (¼ teaspoon) lanolin and 4.0 g (1 teaspoon) beeswax to each muffin cup according to your leader's instructions.

6. Add oil of cinnamon and coconut oil to the muffin cups in the amounts indicated in the following chart.

| Muffin Cup (% Oil of Cinnamon) | Amount of Oil of Cinnamon | Amount of Coconut Oil |
|---|---|---|
| 10 | 1.0 g (¼ teaspoon) | 4.0 g (1 teaspoon) |
| 20 | 2.0 g (½ teaspoon) | 3.0 g (¾ teaspoon) |
| 30 | 3.0 g (¾ teaspoon) | 2.0 g (½ teaspoon) |
| 40 | 4.0 g (1 teaspoon) | 1.0 g (¼ teaspoon) |

7. Before heating the muffin cups, label one end of a set of paper twist ties the same way you labeled the muffin cups (LB, 10, 20, 30, and 40). Label another set of paper twist ties the same way, but add a W to each label (LBW, 10W, 20W, 30W, and 40W). The W stands for water. Also label two paper twist ties that will not be dipped in any test product with 0 and 0W.

← twist tie

# Make and Test Lip Balm

## Leader Guide

- It is okay if a bead has a small portion that is not covered or if one portion is too thickly covered due to a drip. When campers take the beads outside in direct sunlight to evaluate the sunscreening ability, they can observe the parts of the beads that are evenly coated.

- Rather than go outside in direct sunlight, the beads can be placed under a UV lamp in steps 12 and 14. Although it is far easier to evaluate the beads in direct sunlight, the UV lamp can be used at night, during inclement weather, or when it is just not practical for campers to go outside. To prevent eye damage, be sure campers do not look directly at the UV light.

- If taking the group outside is not easy, consider combining steps 12 and 14 together in one outdoor session.

- Campers should throw away the test products made after step 4. Oil of cinnamon is approved as a flavoring in foods and perfumes, but can be irritating to the skin at high concentrations. Do not allow campers to apply these test products to their skin or lips.

**Q1:** Answers will vary, but some campers may say that their lip balm is just like a commercially available product.

**Q2:** The beads that contain the smallest amount of oil of cinnamon turn the darkest.

**Q3:** The test product containing 40% oil of cinnamon is the best sunscreen because its bead is the lightest. (The test product blocked out the most UV radiation.)

**Q4:** The test product has not dissolved off the surface of the beads because the coating can still be seen on the beads. Also, wax is not floating in the water.

---

## CAMPER

twist tie

UV detection bead

8. Thread a UV detection bead on the unlabeled end of each twist tie. Bend the twist tie's end up to hold the bead on the tie as shown. Repeat the procedure for all the twist ties. These beads will be used in steps 10–14.

### Camper Notebook

**Remember...**
Use UV detection beads that turn all the same color.

9. Heat all of the muffin cups (including the LB muffin cup prepared in steps 1–3) on an electric food-warming tray until all ingredients are melted. Stir each mixture occasionally with its own toothpick.

10. Tip the LB muffin cup slightly to maximize the depth of the melted mixture. One at a time, hold the labeled end of the LB and LBW twist ties and completely submerge the beads in the melted mixture. Pull the beads out and allow the coating to harden on the beads (for about 1 minute) before setting the beads down on waxed paper. Repeat this dipping process for the rest of the muffin cups and beads. The O and OW beads will not get dipped.

11. To analyze if your test products are waterproof, place the beads labeled with a W together in a cup of water. The beads should be totally submerged, but the labeled ends should remain out of the water. Let the beads soak for 30–60 minutes.

12. While the beads are soaking in step 11, take the other set of beads outside into direct sunlight to analyze your test products for sunscreen abilities. In the data table at the end of this activity, record the color of each bead. (Beads turn darker as they receive more UV radiation. Lighter beads indicate that the coating blocked the UV radiation. The O bead receives full UV radiation.)

**Q2:** Which beads (besides the O bead) turn the darkest color when exposed to UV light?

**Q3:** Which formulation is best at screening out UV light?

13. After 30–60 minutes, remove the set of beads from the cup of water. Do not dry them off.

**Q4:** Has the test product dissolved off the surface of the beads? How can you tell?

# LEADER

## Make and Test Lip Balm

## Leader Guide

**Q5:** All of the soaked beads turn about the same dark color when exposed to the UV light. Even though the wax is still present in the coating, it appears that the oil of cinnamon is no longer in the coating. The surface of the water that the beads were soaked in has a cinnamon odor. Apparently, the oil of cinnamon migrates through the wax and into the water, indicating that the sunscreen in the lip balm is not waterproof.

**Q6:** None of the test products are waterproof since all of the beads turned dark. None of the test products that were soaked in water blocked out the UV radiation.

**Q7:** Although oil of cinnamon blocks out the sun at high concentrations, it does not make a good sunscreen. Besides the test products not being waterproof, their high oil of cinnamon concentrations might be very irritating to the skin.

## Wrap-Up

Ask campers to discuss their results and their answers to the questions. Be sure campers realize that UV detection beads turn darker as they receive more UV radiation. Point out that, in a real-life situation, either outside water (such as sweat and pool water) or inside water (such as cellular fluids from inside cells) would transport the oil of cinnamon away from the skin's surface, making the test product ineffective in screening out the sun.

---

# CAMPER

### Camper Notebook

14. Repeat step 12 with this second set of beads.

**Q5:** Which of the beads soaked in water (besides the OW bead) turn the darkest color when exposed to UV light?

**Q6:** Which (if any) test product formulation is waterproof? Why or why not?

**Q7:** Would oil of cinnamon make an effective sunscreen? Why or why not?

⚠ *Oil of cinnamon has not been tested for safety in amounts greater than used as a flavor. Therefore, do not apply any of your test products (those products made after step 4) to skin or lips. Discard all of your test products in the trash. The lip balm containing a couple drops of oil of cinnamon (made in steps 1–3 and labeled as LB) can be taken home and used.*

| Muffin Cup | Oil of Cinnamon | Bead Color After UV Exposure (step 12) | Bead Color After Dipping in Water and UV Exposure (step 14) |
|---|---|---|---|
| LB | 2 drops | | |
| 10 | 10% | | |
| 20 | 20% | | |
| 30 | 30% | | |
| 40 | 40% | | |
| 0 | no product | | |

# Leader Guide

# Cover Up, Screen, or Block?

> ✋ **Suggestions for leading this activity begin on the next page.**

By using UV detection beads, campers learn how certain sun products, clothing, and sunglasses can reduce exposure to the sun's ultraviolet (UV) radiation.

## Some Leader Background Information

Sunblocks and sunscreens are labeled with a sun protection factor (SPF). The SPF number tells people about how long they can stay in the sun without burning. An SPF 15 sunscreen lets people stay in the sun 15 times longer without burning than they normally could without sunscreen. For example, if it usually takes someone only 20 minutes to burn without sunscreen, then when that person wears an SPF 15 sunscreen he or she can stay in the sun for 15 (the SPF number) × 20 minutes = 300 minutes (5 hours) without burning. Campers should know, though, that the U.S. Food and Drug Administration (FDA) is not sure this formula holds up for SPF numbers above 15. Products that claim SPFs greater than 30 are not necessarily more than twice as powerful at protecting people as SPF 15 sunscreens. However, some researchers disagree with the FDA.

In this activity, UV detection beads are covered with sun protection products, fabric, and sunglasses to analyze the effectiveness of these materials against UV radiation. Campers observe that the beads turn a darker color when they are exposed to more UV radiation. As campers might expect, the SPF ratings of the products correlate with how quickly and how deeply the beads change shade. Beads covered with no sun protection product or low SPF product quickly change to a deep shade, while those covered with a maximum protection (SPF 30 or higher) product remain white or nearly white. Beads covered with intermediate levels of SPF show a change somewhere in between.

Depending on the fabrics tested, campers may notice some difference in how each fabric protects the bead from UV radiation. Different types of clothing can offer different degrees of protection from the sun. In terms of UV protection, the density of the weave plays a more important role than the color or type of fabric.

Today, most plastic sunglass lenses are treated with a coating or additive to block UV light. Although probably made of a similar plastic material, the cup lacks this coating or additive. Campers should notice a difference in the degree of color change between beads covered by lenses with different ratings. Those with a higher rating should block more UV than those with a lower rating or no specific rating. Sunglasses with a 100% UV rating should block all of the UV rays. (Some UV light may reach the beads indirectly if there are gaps around the edge of the lens.)

Make sure campers know there are many different ways to protect themselves from UV radiation. Studies show that almost one-third of the total daily UV hits the earth between 11 AM and 1 PM. People should try and plan their time outdoors before or after these hours. A useful rule to remember is that when the length of a person's shadow is shorter than his or her height, the risk of getting sunburned is high. When going outside during these times, stay in the shade whenever possible. Also, don't forget to cover up. Most clothing is a good blocker of UV radiation, but some types are better than others. Wearing a hat is an excellent way to protect the face, neck, and ears. Wear sunglasses that provide 99–100% UVA/UVB protection. Lastly, always use sunscreen. Dermatologists recommend that people of all skin shades regularly use at least SPF 15 sunscreen. People should reapply sunscreen every 2 hours as long as they are outdoors.

# LEADER

## Cover Up, Screen, or Block?

### Leader Guide

### Materials for Camper Procedure—See Left

### Materials for Getting Ready

✓ UV detection beads

☞ *UV beads are available from science and teacher supply stores or from similar sources on the Internet. Packages of all one color are preferred for this activity, because each set of test beads has to be the same color. (Avoid using yellow beads.) You may purchase packages of assorted colors, but then you'll have to sort the beads into like colors. (See step 1 of Getting Ready.)*

✓ black construction paper

✓ scissors

✓ glue

✓ 3 different types of fabric

☞ *The most important variable in fabric UV protection is the tightness of the weave. To ensure that campers see a visible difference in results, select one fabric with tight weave and one fabric with loose weave. The third fabric's weave can be whatever you choose. You may want to test your fabrics in advance.*

✓ ruler

✓ permanent marker

✓ masking tape or self-stick label

✓ 2 types of sunglasses with different UV ratings

☞ *For the last several years, sunglasses sold in the United States have at least some degree of UV protection. To ensure that campers see a visible difference in results, be sure to select one type of sunglasses with 100% protection and the other type of sunglasses with less protection. Those with 100% protection will typically be labeled as such, while those with lesser protection may not be marked. You may have to pick a few unmarked pairs and test them to find sunglasses with less than 100% protection.*

✓ very small screwdriver to remove sunglass lenses from their frames

✓ (optional) UV light

☞ *Light should not be directed into the eyes.*

Page 248

---

## CAMPER

### Cover Up, Screen, or Block?

#### Overview

You've probably heard that using sunscreen and wearing proper clothing and sunglasses will protect you from the sun. In the following activity, you will test the protection of different sunblocks, sunscreens, clothing, and sunglasses using UV detection beads.

#### Materials

✓ 3 bead setups (prepared by leader)
✓ chalk
✓ gallon zipper-type plastic bag
✓ 4 sun protection products (such as sunscreens and sunblocks) having a wide range of SPF ratings
✓ cotton swabs
✓ 3 fabric squares with different weaves (prepared by leader)
✓ tape
✓ clear plastic cup
✓ 2 sunglass lenses with different UV ratings (prepared by leader)

#### Procedure

**Test of Sun Protection Products**

1. As shown at the left, use chalk to label the paper next to each bead in the five-bead setup with the SPF rating of the sun protection product you are going to test. One bead will have no sun protection product (0 SPF) and will be the control. Slide the construction paper into the gallon-sized plastic bag.

2. Use a clean cotton swab for each sun protection product in this step. Smear a small amount of the appropriate product on the plastic bag directly over each bead. It is important to apply the same amount of product evenly over each bead. Use the cotton swab to spread the sun protection product into a circle that is about 1½ inches (about 4 cm) in diameter. Circles of this size should provide protection to the tops and sides of the beads.

3. Write down the name and SPF of each sun protection product and the starting color intensity of each bead in the data table at the end of this activity. Also record the time of day and weather conditions (such as sunny, partly sunny, or cloudy).

Page 119

# Cover Up, Screen, or Block?

## Leader Guide

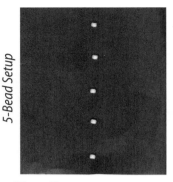
*5-Bead Setup*

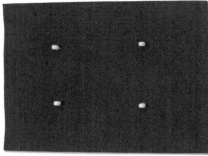
*4-Bead Setup*

### Getting Ready

1. Decide on how many sets of test beads you want to prepare. Ideally, each camper or group should get one five-bead setup (for testing the sun protective products) and two four-bead setups (for testing the fabrics and sunglass lenses). Each set should contain beads of the same color. If your package of beads is an assortment of colors, use a UV light or take the beads out in the sun to clearly view the colors of the beads. (Each bead set needs to be the same color because different-colored beads respond to the sun's UV light at different rates and with different degrees of color change. Avoid selecting yellow beads because their shade changes are the hardest to observe.)

2. For each five-bead setup, cut a piece of black paper so it fits in a gallon-sized plastic bag. Evenly space and glue five beads of the same color on the black paper as shown, one bead for each sun protection product to be tested and one bead for the control (no sun protection product). Make sure not to get glue on top of the beads. Let the glue dry.

3. For each four-bead setup, glue four same-color beads to the piece of black paper as shown. Avoid getting glue on top of the beads. Let the glue dry.

4. Assign each type of fabric a number. Cut the fabric into 3-inch × 3-inch (8-cm × 8-cm) squares and label each square with its appropriate number using a permanent marker. (If you can't write directly on the fabric, attach a small piece of masking tape or a self-stick label. Place the number in the lower corner of the fabric.)

---

## Camper Notebook

4. Take the setup outside in the direct sunlight. Without removing the plastic bag, immediately observe and record the color intensity of the beads (such as white, nearly white, light, medium, or dark). If you can't determine the color intensity of the beads through the sunscreens, take the beads indoors, open the plastic bag, and immediately observe the color intensity of the beads (so that the color does not significantly fade).

**Q1:** Compare the color intensity changes of the beads with the SPF ratings of the products covering them. What is the trend?

**Q2:** Imagine doing this test during a commercial to sell sunscreen. Do you think this test would convince people to buy one particular sun protection product over another? Explain your answer.

### Test of Fabric and Sunglasses

5. Tape each piece of fabric over a bead in a setup as shown at left. Use only one piece of tape for each piece of fabric. The fourth bead will serve as the control. You will leave this bead uncovered to expose it to direct sunlight.

6. With tape, secure the plastic cup and sunglass lenses over the beads in another setup as shown at lower left. The fourth bead will serve as the control.

7. Record the starting color intensity of each bead in the data table. (You can flip up the objects to look at the beads but be sure to re-cover the beads after checking their color intensity.) Record the UV protection rating of the sunglass lenses if you know it. Also record the time of day and weather conditions (such as sunny, partly sunny, or cloudy).

**Important...** Do not peek at the beads under the objects until you've brought the setups back inside.

8. Carry the setups outside into the sun. After the control beads turn dark (usually in 1 or 2 minutes), bring the setups indoors and immediately look under the covers to observe the color of the beads. Record the bead color intensity (such as white, nearly white, light, medium, or dark) in the data table.

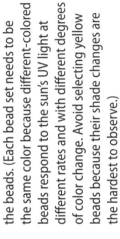
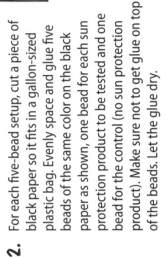
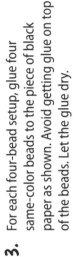

# LEADER

## Cover Up, Screen, or Block?

### Leader Guide

Make enough fabric squares so that each camper or group gets one square of each fabric type.

5. Assign each type of sunglasses a number. Use a small screwdriver to remove the lenses from each pair of sunglasses. Label each lens with its appropriate number using a small piece of masking tape or self-stick label. (Place the label at the very edge of the lens.) Prepare enough lenses so each camper or group gets one lens of each type.

6. As an alternative to using UV detection beads, punch holes in black construction paper (one hole for each bead location). Tape a sheet of sun-sensitive paper under the construction paper so that the photosensitive side is up. Campers should then do each step in the camper procedure. The sun-sensitive paper changes very quickly, so expose it to sunlight for only about 20 seconds. Then bring the paper into a dark room to develop the paper according to its directions. Once developed and dry, check its shade at each of the hole locations.

### Introducing the Activity

Tell campers that, since humans can't sense UV radiation, they will use UV detection beads to experiment with and learn about products that can be used to reduce exposure to UV radiation. Show campers how the UV detection beads work by putting the beads under a UV light or taking the beads out in the sun. Be sure to include the same color of beads you glued to the black paper in Getting Ready. Tell campers that the beads change color intensity when exposed to UV radiation.

### Procedure Notes and Tips

- To save time and materials, you may want to have some campers test the sun protection products (steps 1–4), some test the fabrics (steps 5, 7, 8), and some test the sunglass lenses (steps 5, 6, 8).

- Be sure to include sun protection products with SPF ratings of 8 or lower.

---

## CAMPER

### Cover Up, Screen, or Block?

**Camper Notebook**

**Q3:** Based on your results, what type of clothing do you think would best protect you from the sun? What is your evidence?

**Q4:** Compare the UV protection of the sunglasses to the UV protection of the cup. Why are they different?

**Test of Sun Protection Products**

| Name of Product | SPF | UV Bead Color Intensity | |
| --- | --- | --- | --- |
| | | Start | After Sun Exposure |
| no sunscreen | 0 | | |
| | | | |
| | | | |
| | | | |

Time of day and weather conditions:

**Test of Fabric and Sunglasses**

| | Type of Cover | UV Protection Rating | UV Bead Color Intensity | |
| --- | --- | --- | --- | --- |
| | | | Start | After 1–2 Minutes |
| Clothing Test | control (direct sun) | 0 | | |
| | fabric sample 1 | — | | |
| | fabric sample 2 | — | | |
| | fabric sample 3 | — | | |
| Sunglasses Test | control (direct sun) | 0 | | |
| | sunglasses brand 1 | — | | |
| | sunglasses brand 2 | — | | |
| | cup | — | | |

Time of day and weather conditions:

## Cover Up, Screen, or Block?

### Leader Guide

- Some types of UV beads may start to change color even under fluorescent lights. That's okay. These changes are typically not as intense as when the beads are exposed to UV light.

**Q1:** There should be a general trend from low SPF (darker bead color) to high SPF (lighter bead color).

**Q2:** This experiment might help to sell sun protection products because buyers can see the shade changes of the UV beads. A bead protected with a sun protection product stays lighter than an untreated bead, showing that exposure to UV radiation is being decreased by the product.

**Q3:** Answers will vary depending on the types of fabric tested. Generally, tighter woven fabrics block more UV radiation than looser woven fabrics.

**Q4:** Answers will vary depending on the types of sunglasses tested. Sunglasses having a higher UV rating should block more UV than those with a lower rating or no specific rating.

### Wrap-Up

Have campers share their results with the group. Discuss the factors that affect how much radiation reaches the earth's surface. (For help with the discussion, see Leader Background Information.) Review preventative measures that can be taken to help prevent skin damage from UV radiation. (For help with the review, see below.)

---

#### How to Reduce Exposure to UV Radiation

- Plan outdoor activities before 11 AM and after 1 PM.
- Try to stay in the shade.
- Wear clothes to cover up.
- Wear a hat with a brim measuring 2–3 inches all the way around.
- Wear sunglasses having 99–100% UVA/UVB protection.
- Always use sunscreen or sunblock.

# LEADER

## Sunning Straws

### Leader Guide

> Suggestions for leading this activity begin on the next page.

Sunblocks and sunscreens work by either blocking or absorbing ultraviolet (UV) radiation from the sun. In this activity, campers use benzophenone and a UV light reaction to quantitatively measure the amount of UV protection that different sun protection products provide.

### Some Leader Background Information

Sunblocks block UV rays before they hit your skin. Sunblocks contain minerals that reflect UV light. Zinc oxide is a common sunblock. Lifeguards are sometimes seen with this white paste on their noses. Sunscreens, on the other hand, contain chemicals that absorb UV radiation and convert it into heat. Sunscreens are the oily and creamy lotions (with brand names such as Coppertone® and Hawaiian Tropic®) that most of us are familiar with. Active ingredients in sunscreens usually include names such as salicylate, cinnamate, and benzophenone. The best sunscreens also contain some sunblocking agents such as very tiny particles of zinc oxide or titanium oxide.

Sunblocks and sunscreens are labeled with a sun protection factor (SPF). The SPF number tells you about how long you can stay in the sun without burning. An SPF 15 sunscreen lets you stay in the sun 15 times longer without burning than you normally could without sunscreen. For example, if it usually takes you only 20 minutes to burn without sunscreen, then when you wear an SPF 15 sunscreen you can stay in the sun for 15 (the SPF number) × 20 minutes = 300 minutes (5 hours) without burning. You should know, though, that the U.S. Food and Drug Administration (FDA) is not sure this formula holds up for SPF numbers above 15. Products that claim SPFs greater than 30 are not necessarily more than twice as powerful at protecting you as SPF 15 sunscreens. However, some researchers disagree with the FDA.

In this activity, benzophenone is used to quantitatively measure the UV protection of different sun protection products. Benzophenone is a ketone with phenyl rings attached. It reacts with UV light to form pinacol. The other component in the straw (isopropyl alcohol) ends up forming acetone (another ketone), which is a liquid and stays in solution. Since acetone does not have the phenyl groups attached to the ketone, it does not react with UV light. The overall reaction is:

2 benzophenone + 2 isopropyl alcohol + UV light → benzopinacol + 2 acetone

Many of the compounds in sunscreens contain some type of carbonyl bond that is attached to a phenyl ring or other conjugated system. In the straws, the benzophenone acts like a sunscreen compound. When it absorbs UV light, it gives a solid pinacol that is observed as the white precipitate in the straw. Sunscreens work by absorbing the UV light before it hits our skin. Light energy from the sun reacts with the sunscreen and converts to heat, dissipating harmlessly.

# Sunning Straws

## Leader Guide

### Materials for Camper Procedure—See Left

### Materials for Getting Ready

- ✓ benzophenone
  ⚠ *Wear goggles when handling benzophenone and the benzophenone solution. Benzophenone is available from chemical supply companies.*
- ✓ balance
- ✓ isopropyl alcohol
  ☞ *Be sure to use pure or 99% isopropyl alcohol and not 70% or 90%. Try looking in chemical supply companies, hardware stores, or drugstores.*
- ✓ graduated cylinder
- ✓ container to hold the benzophenone solution
- ✓ transparent, colorless polyethylene (plastic) drinking straws
- ✓ hot-melt glue gun and glue
- ✓ disposable pipet
- ✓ narrow container to hold straws upright
- ✓ (optional) pliers

### Getting Ready

1. Figure out how many test straws are needed. (Each setup requires one straw for each sun product to be tested, plus a straw for the control.) Each straw holds about 5 mL solution. Prepare the benzophenone solution in a ratio of 1 g benzophenone to 20 mL isopropyl alcohol.

2. Seal one end of each straw with hot-melt glue. Be sure to either completely plug the straw's hole with glue, or fold the straw about ¼ inch (0.5 cm) from the end and hot glue the fold down to seal the end closed. (Pliers can be used to hold the fold down until glue sets.)

3. Use a disposable pipet to add some benzophenone solution to each straw. Make sure the solution does not leak out of the bottom seal, then continue filling the straw to within ¾ inch (2 cm) from the top of the straw. Hot glue the top of the straw. Be sure to check for leaks.

---

# Sunning Straws

### Overview

You've probably heard commercials for sunblocks and sunscreens talk about how their products block and absorb ultraviolet (UV) radiation. In this activity, you will actually measure the amount of UV protection that different products offer.

### Materials

- ✓ straws containing a light-detecting chemical (prepared by leader)
  ⚠ *Wear goggles when handling the straws.*
- ✓ permanent marker
- ✓ sunblocks and sunscreens having different sun protection factor (SPF) numbers
- ✓ cotton swabs
- ✓ cardboard or plastic tray to hold the straws horizontally
- ✓ aluminum foil
- ✓ facial tissues or paper towels
- ✓ narrow container (such as a tall cup) to hold the straws upright
- ✓ metric ruler

### Procedure

1. Designate one straw for each sunblock and sunscreen product to be tested by numbering the top of each straw and writing the product names in the data table at the end of this activity. Label one straw "C" for control. This straw will be left as is.

2. Use a cotton swab to spread each sun protection product on the entire outside surface of the appropriate straw. Be sure to use a different cotton swab for each product. Place the straws horizontally on a tray covered with aluminum foil, leaving as much space between the straws as you can.

3. Take the tray outside and expose the straws to direct sunlight for at least 90 minutes.

4. After exposing the straws to the sun, bring the straws inside and wipe the sun protection products off the outside using facial tissues or paper towels.

# LEADER

## Sunning Straws

**Leader Guide**

### Introducing the Activity

Discuss how sunblocks and sunscreens work. Be sure to talk about SPF. (For help with the discussion, see Leader Background Information.) Explain to campers what they will be doing in this investigation.

### Procedure Notes and Tips

- You may wish to have campers make the straws by doing Getting Ready steps 2 and 3, or just step 3. Adult supervision is required when using the hot-melt glue gun. Be sure students wear goggles when handling the benzophenone solution.

- Make sure campers use a different cotton swab for each sun protection product. You may want to keep one cotton swab with each product so it can be used by all campers.

- When placing the tray of straws outside, it may be necessary to weigh it down with a rock or two to prevent the wind from blowing the experiment away.

- To ensure that all of the solid has settled in the straws during step 5, you may want to have campers take the measurements twice in step 6.

**Q1:** After sun exposure, the control straw has white solid inside.

**Q2:** The more UV radiation that the solution inside the straw was exposed to, the more white solid forms inside the straw. Therefore, less white solid means better UV protection.

**Q3:** The product with the highest SPF should generally offer the best protection against UV radiation.

### Wrap-Up

Have campers share their results. Discuss the chemical reaction caused by the sun when benzophenone converts to benzopinacol. (For help with the discussion, see Leader Background Information.) Be sure campers understand that more white solid in the straw means that more UV radiation went through the straw to cause the chemical reaction.

---

# CAMPER

**Camper Notebook**

**Q1:** What do you see in the straw that had nothing applied on the outside (the control)?

**5.** Look at each straw. If any solid is present in the straw, hold it upright and gently tap the side of the straw to allow the solid to settle to the bottom of the straw. The solid is called benzopinacol. Store the straws upright in the narrow container.

**6.** After allowing the solid to settle for a few minutes, use a metric ruler to measure the height of the white product present in each straw. Record the heights in the data table.

**Q2:** How is this procedure a quantitative measure of the amount of UV radiation hitting the solution inside each straw?

**Q3:** Which product offers the best protection against UV radiation?

| Sunblock or Sunscreen Product | Height of White Solid |
|---|---|
| C    no product (control) | |
| 1 | |
| 2 | |
| 3 | |
| 4 | |
| 5 | |
| 6 | |
| 7 | |

**FYI...**
By answering these questions, campers should discover that the reaction of UV light and benzophenone in isopropyl alcohol produces a white precipitate. If campers have trouble with this concept, try asking them to predict what would happen if a straw system was exposed to a UV light. Then, expose the straw system and have students observe the precipitate that formed.

**Leader Guide**

# Suntan in a Bottle

> 👉 **Suggestions for leading this activity begin on the next page.**

In this activity, campers apply a sunless tanning product to various fabric samples containing natural and synthetic fibers to observe how the product works.

## Some Leader Background Information

The concept of tanning lotions that give the skin an artificial tan is actually quite old. In 1960, Coppertone® introduced its QT® or Quick Tanning Lotion. The product was not very popular because it usually made a person look orange, rather than brown. It commonly left streaks and splotches on the skin, and if you were not careful to wash your hands after applying it, it would turn your palms orange too! Some customers also did not like the lotion's odor. Since then, the science of sunless tanning has become more advanced. Today, sunless tanning lotions give a more natural color, and (with the right equipment) can be applied very evenly to the skin.

According to the American Academy of Dermatology, the best sunless tanning lotions are ones that have dihydroxyacetone (DHA) as the active ingredient. DHA is a colorless substance, usually refined from sugar beets or sugar cane. DHA works by interacting with proteins in the outermost layer of the skin (the epidermis) to produce a color change that resembles a tan. Products that work this way are sometimes called extenders.

Extenders are sold as gels, lotions, and sprays. You can buy many of them over the counter for use at home. However, more and more people are going to salons to have their tans professionally sprayed on. The two most popular technologies are airbrush tanning systems and automatic spray-on booths.

If you go to a salon for a sunless tan, The U.S. Food and Drug Administration (FDA) recommends that you make sure the salon has a way to protect your eyes, lips, and other sensitive areas from the spray. The salon should also have safeguards to prevent you from inhaling or swallowing the spray.

Extenders take about 45 minutes to an hour to begin taking effect. The complete color change may take up to a day. The tan usually fades within a week because the outer layer of your skin is constantly shedding. To keep your tan, you need to reapply the sunless-tanning product every few days.

Extenders are generally safe to use, but some people can be allergic to the ingredients. Several days before using these products, you should apply a tiny amount to a small portion of your skin to see if you have an allergic reaction.

After you applied the sunless tanning product to the fabrics in this activity, you probably didn't notice any color change for an hour or more. Maximum color develops over the course of several hours. These same effects are seen on human skin.

Sunless tanning is not true tanning. The tanning effect is produced by the reaction of DHA with the proteins (keratins) found naturally in skin. The reaction produces chemicals called "melanoids" that are similar in color to melanin but rather different chemically. Proteins are also found in other animal products such as wool and silk. Both of these fabrics and the synthetic fiber nylon are all examples of polyamides, which will react with the DHA in sunless tanning products and cause these fabrics to change color. Other natural fabrics (such as cotton) and synthetic fabrics (such as polyester) do not react with DHA and therefore do not become noticeably tan.

# LEADER

## Suntan in a Bottle

# Leader Guide

## Materials for Camper Procedure—See Left

### Introducing the Activity

Discuss the warnings listed in the Camper Notebook and on the bottle of sunless tanning product.

### Procedure Notes and Tips

- Make sure campers apply the sunless tanning product evenly over the entire piece of fabric.
- As a variation, different campers may use different sunless tanning products and then compare results.

**Q1:** Answers will vary depending on the product used.

---

# CAMPER

## Suntan in a Bottle

### Overview

Sunless tanning products are a popular way for people to appear tan without exposing themselves to harmful UV rays. In this activity, you'll test a sunless tanning product and observe how it affects different kinds of fabric.

### Materials

- newspaper
- different types of white fabric (such as cotton, wool, polyester, nylon, and silk)
- scissors
- glue
- sunless tanning product containing dihydroxyacetone (DHA)
- ⚠ Read the safety precautions on the product label. Be careful not to get the product on your clothes. Wash your hands thoroughly after applying the product.
- cotton swab

### Procedure

1. Lay down newspaper to protect the work area. Cut two squares from each fabric, each measuring about 1½ inches × 1½ inches (about 4 cm × 4 cm).
2. Glue the two squares of each fabric type in the data table at the end of this activity. (One square will be untreated, serving as the control, and the other will be treated with a sunless tanning product.) In the data table, record the types of fabric. (It's okay if you don't know that information.) Let the glue dry.
3. Carefully use a cotton swab to evenly apply sunless tanning product to each square of fabric in the "Treated Sample" column of the data table. Notice what time it is. Thoroughly wash your hands when you are done.

**Q1:** What is the color of the sunless tanning product when it is first applied?

# Suntan in a Bottle

## Leader Guide

**Q2:** Wool, silk, and nylon become light tan.

**Q3:** Fabrics made from animal products such as wool and silk act like human skin and become light tan. (These fabrics are examples of polyamides, which are similar to proteins found naturally in skin.) The synthetic fiber nylon is also a polyamide and becomes light tan.

## Wrap-Up

Discuss how sunless tanning products work to change the color of skin. (For help with the discussion, see Leader Background Information.)

---

## Camper Notebook

4. Observe the results after several hours. Record the total treatment time and your observations in the data table.

**Q2:** Which fabrics react with the sunless tanning product and become "tanned"?

**Q3:** What do the "tanned" fabrics have in common? (Hint: Think about what each fabric is made of.)

# LEADER

## Suntan in a Bottle

### Leader Guide

---

**CAMPER**

## Camper Notebook

| Type of Fabric | Fabric Treated for ___ Hours and ___ Minutes | | Observations |
|---|---|---|---|
| | Untreated Sample (Control) | Treated Sample | |
| | Glue Fabric Here | Glue Fabric Here | |
| | Glue Fabric Here | Glue Fabric Here | |
| | Glue Fabric Here | Glue Fabric Here | |
| | Glue Fabric Here | Glue Fabric Here | |
| | Glue Fabric Here | Glue Fabric Here | |

Healthy Skin—Suntan in a Bottle

**Leader Guide**

# A Look at Bleaching

> **Suggestions for leading this activity begin on the next page.**

Despite warnings from dermatologists and health professionals about the possible risks of skin bleaching, millions of dark-skinned people around the world continue to use a variety of creams, lotions, and soaps to try to achieve a lighter complexion. Some of these products contain ingredients that can have very harmful effects on the skin. In this activity, campers use fabrics to observe the effects of common household and skin care products containing bleaching agents.

## Some Leader Background Information

The most common ingredient used for skin bleaching is hydroquinone (HQ). Creams, lotions, and other cosmetics containing 2% HQ are available over the counter in the United States. These products are effective for fading freckles and age spots and are generally considered safe if used as directed. In the United States, products containing higher concentrations of HQ are prescribed by doctors to treat more serious skin conditions.

Legal over-the-counter and prescription products containing HQ are intended for short-term use on a limited area of the skin. Problems arise when people misuse these products by applying them to larger areas of skin for months or years at a time. Some black-market products contain much higher HQ levels than are present even in prescription-strength products. One illegal cream sold in South Africa contained 25% HQ.

HQ can have serious side effects when misused. It can cause skin irritation, and users often appear sunburned because their faces turn red and puffy. With continued use, the skin can become dark and blotchy, leaving people with a worse complexion than when they started. Some scientists suspect that HQ can cause cancer.

The science behind this activity does not directly represent the science of skin-bleaching products. Even so, this activity shows that some chemicals can bleach pigments and sometimes damage fabric. Most likely campers noticed that some fabrics became lighter, and some fabric fibers may have started to break down.

Many household products contain small amounts of bleach. Household bleach is a solution of sodium hypochlorite. Dishwasher detergents and laundry bleach pens contain sodium hypochlorite in lower concentrations than what is found in household bleach. Anti-aging cream contains hydroquinone, the same ingredient found in skin-bleaching creams. Some acne-care products contain benzoyl peroxide and some oxygen cleaners contain sodium percarbonate. Tooth whiteners often contain hydrogen peroxide.

The results of this activity show that the tea-dyed fabric is bleached by most (if not all) of the products tested. Some of the cotton fabrics are bleached by the dishwasher detergent. Since many of the other fabrics are dyed by the manufacturer to be colorfast, most of the test products do not affect these fabrics very much.

Healthy Skin—A Look at Bleaching

# LEADER

## A Look at Bleaching

# Leader Guide

## Materials for Camper Procedure—See Left

## Materials for Getting Ready

- ✓ white cotton fabric (either woven or knit like T-shirt fabric)
- ✓ scissors capable of cutting fabric
- ✓ ruler
- ✓ regular tea bags
- ✓ water
- ✓ glass beaker or stainless steel pan
- ✓ hot plate or stove
- ✓ paper towels
- ✓ (optional) materials to demonstrate the damaging effects of bleach
  - samples of the dark fabric used in the camper procedure
  - plastic cup for each fabric sample
  - household liquid chlorine bleach
  - craft stick
  - goggles, rubber gloves, and apron

## Getting Ready

1. Cut the white cotton fabric into rectangles (one per camper) measuring about 6 inches × 8 inches (15 cm × 20 cm).
2. Decide how much water is needed to dye the white fabric. Add two tea bags for every cup of water. Heat the water to boiling. Allow the tea bags to steep until a very strong tea is made. Remove the tea bags.
3. Completely submerge the white fabric rectangles in the strong tea. Allow the fabric to remain in the strong tea at least 15 minutes.
4. Rinse the fabric with water and wring out the excess liquid. Allow the fabric to air dry.

Healthy Skin—A Look at Bleaching

Page 260

---

# CAMPER

## A Look at Bleaching

### Overview

Some people think they look better if they are suntanned, while other people want their skin to be lighter. Either tanning or skin bleaching to change skin color is risky. In this activity, you'll get an idea of the risks of skin bleaching by observing the effects of different household products on fabric samples. (Skin testing will not be done for safety reasons.)

### Materials

- ✓ large plastic garbage bag
- ✓ tea-dyed fabric (prepared by leader)
- ✓ several dark-colored fabric samples (such as black cotton broadcloth, black or blue cotton denim, black wool, black nylon, and black polyester)
- ✓ scissors capable of cutting fabric
- ✓ ruler
- ✓ rubber gloves
- ✓ up to 6 household and skin care products such as:
  - liquid or gel dishwasher detergent
  - laundry bleach pen
  - anti-aging cream containing hydroquinone
  - acne care product containing benzoyl peroxide
  - "oxygen" cleaner (such as Oxiclean™ or OxiMagic™)
  - tooth whitening gel
  - hydrogen peroxide
- ⚠ *Do not use undiluted liquid household laundry bleach in this activity.*
- ✓ applicators for each product such as:
  - cotton swab (such as Q-Tip®)
  - small sponge piece no larger than 1 inch × 1 inch (2 cm × 2 cm)
- ✓ hand soap

### Procedure

1. Cover your work area with a large plastic bag to prevent damage to the surface if test substances soak through the fabrics.

Healthy Skin—A Look at Bleaching

Page 127

# A Look at Bleaching

## Leader Guide

5. You may want to expose a set of fabric samples to full-strength household liquid chlorine bleach for 24 hours to demonstrate its damaging effects. For safety reasons, do this in an area away from the campers. Show campers the bleached fabric samples in Wrap-Up.

- Cut about 1 inch × 1 inch (3 cm × 3 cm) squares from the same fabrics the campers are going to test. Place each fabric into a separate plastic cup. Label each cup with the type of fabric being tested.

- Put on goggles and add bleach to the cups until the squares are entirely submerged. Poke the squares down with a craft stick if necessary.

⚠ *Wear goggles while pouring bleach and observing samples. Avoid skin and eye contact with the bleach. If contact occurs, follow the safety instructions on the bottle label. Wash hands after handling bleach, containers, and craft stick.*

- Allow the fabric squares to remain in the bleach undisturbed for 24 hours. When time is up, place each cup containing bleach and fabric under a faucet of gently running water to dilute the bleach. (Make sure the fabric does not go down the drain.) Flush with plenty of water. Finally, pat the fabric squares dry between pieces of paper towel.

## Introducing the Activity

Discuss the various products that are available to bleach skin, fabric, dishes, teeth, and other items. (For help with the discussion, see Leader Background Information.) Tell campers what products they will be testing on their fabric. If you want, identify the bleaching agent in each product. Discuss the safety issues related to each product. (For help with the discussion, see the product labels.)

---

### Camper Notebook

2. If not done already, cut each fabric sample into a large rectangle about 6 inch × 8 inch (15 cm × 20 cm). Mark one edge of each piece as the top by cutting off both corners as shown in the figure at left. (You'll need to know which edge is the top in order to keep track of where you've placed the test products on the fabrics.)

3. Use the figure at left to record where you will place each product to be tested. Be sure to evenly distribute the locations, allowing at least a 2-inch × 2-inch (5-cm × 5-cm) area for each product. Record the name of each product and its bleaching agent on the figure.
   👉 *Read the product labels or ask the leader to identify the bleaching agents. Examples include sodium hypochlorite, hydroquinone, benzoyl peroxide, sodium percarbonate, and hydrogen peroxide.*

4. Put on rubber gloves. Use a swab or sponge applicator to apply each product to each fabric. A mark no bigger than about 1 inch × 1 inch (1.5 cm × 1.5 cm) works well because some products will spread out on the fabric over time. Be sure to use a different applicator for each product.
   👉 *Follow any safety precautions listed on the label when handling each product.*

5. Allow the products to remain on the fabrics for at least 30 minutes. When time is up, put on rubber gloves and rinse the fabrics. Then, wash the fabrics with hand soap to remove any oily residue. Squeeze out as much water as possible. Roll the fabric samples in paper towels and twist the paper towels to remove excess moisture.

6. Record each type of fabric, each test product, and your observations in the data table at the end of this activity.

**Q1:** Generally, what do the household products do to the fabrics?

**Q2:** Which fabrics are most sensitive to the bleaching agents in the different products?

# LEADER

## A Look at Bleaching

**Leader Guide**

### Procedure Notes and Tips

- To save camp time, you may want to cut each of the dark-colored fabric samples into 6-inch × 8-inch (15-cm × 20 cm) rectangles ahead of time.

- You may want campers to make their own tea-dyed fabric. If so, instruct campers on the procedure as outlined in steps 1–4 of Getting Ready. Be sure to allow time for the fabric to dry.

- Make sure campers use different applicators for each product. You may want to keep one applicator with each product so it can be used by all campers.

- The untreated parts of each fabric sample used in the camper procedure serve as the control. Campers can look at these parts of the fabric when making comparisons during their observations in step 6 of the camper procedure.

**Q1:** Some fabrics become lighter and, in some cases where there was concentrated exposure, the fibers begin to break down.

**Q2:** The tea-dyed fabric is bleached by most (if not all) of the products tested. Some of the cotton fabrics are bleached by the dishwasher detergent.

### Wrap-Up

Ask camper to share their results. If you treated fabric samples with full-strength household liquid chlorine bleach (as outlined in step 5 of Getting Ready), show campers your results. Discuss the damaging effects of skin bleaching. (For help with the discussion, see Leader Background Information.)

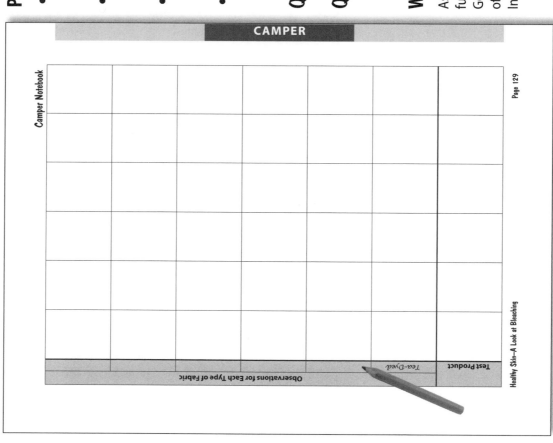

Healthy Skin—A Look at Bleaching

Page 262

# Leader Guide

# UV Detective Challenge

☞ **Permission is granted for you to copy and distribute this take-home activity for your event.** (The master is provided in the Camper Notebook.)

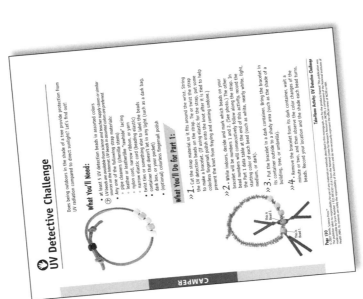

LEADER

# References

## Disease Control Is in Your Hands

American Academy of Child & Adolescent Psychiatry Website. Publications. Facts for Families. Obsessive-Compulsive Disorder in Children and Adolescents. www.aacap.org (accessed May 3, 2006).

American Association for the Advancement of Science Website. Eureka Alert! www.eurekalert.org/pub_releases/2003-09/asfm-aua091103.php (accessed May 12, 2006).

American Museum of Natural History Website. Epidemic! The World of Infectious Disease. www.amnh.org/epidemic (accessed April 26, 2006).

American Society for Microbiology Website. Experiment. Let's Get Small. www.microbe.org (accessed April 24, 2006).

American Society for Microbiology Website. Microbes Are the Foundation of Life. www.microbe.org/microbes/mysteries.asp (accessed April 24, 2006).

American Society for Microbiology Website. Types of Microbes. www.microbeworld.org/htm/aboutmicro/microbes/types_start.htm (accessed April 24, 2006).

Association of Christian Teachers and Schools Website. School-Wide Handwashing Campaigns Cut Germs, Absenteeism. acts.ag.org (accessed April 26, 2006).

Centers for Disease Control and Prevention Website. A–Z Index. SARS. What Everyone Should Know. Basic Information About SARS. www.cdc.gov (accessed July 14, 2006).

Centers for Disease Control and Prevention Website. Emerging Infectious Diseases. Antibacterial Household Products: Cause for Concern. www.cdc.gov (accessed July 14, 2006).

Centers for Disease Control and Prevention Website. Hand Hygiene Guidelines Fact Sheet. www.cdc.gov/od/oc/media/pressrel/fs021025.htm (accessed April 24, 2006).

Centers for Disease Control and Prevention Website. National Center for Infectious Diseases. Food-Related Diseases. www.cdc.gov/ncidod/diseases/food/index.htm (accessed May 12, 2006).

Centers for Disease Control and Prevention Website. Press Room. Press Releases. July 14, 2005. New Study Demonstrates Simple Handwashing with Soap Can Save Children's Lives. www.cdc.gov/od (accessed April 26, 2002).

The Chicago Tribune Website. Infection Epidemic Carves Deadly Path. www.chicagotribune.com/news/specials/chi-0207210272jul21,1,4897794.story (accessed May 11, 2006).

The Chicago Tribune Website. Lax Procedures Put Infants at High Risk: Simple Actions by Hospital Workers, Such as Diligent Hand Washing, Could Cut the Number of Fatal Infections. www.chicagotribune.com/news/specials/chi-0207220180jul22,1,4766722.story (accessed May 11, 2006).

FYI... These sources were used to compile the information in this book. Visit www.terrificscience.org/thrive/ for updates on these topics.

The Chicago Tribune Website. State Falls Short in Tracking Diseases. www.chicagotribune.com/news/showcase/chi-0207230227jul23,1,6461469.story (accessed May 11, 2006).

CNN.com Website. Dr. Sanjay Gupta: Alarm About Hospital Germs. www.cnn.com (accessed April 24, 2006).

Curtis V.; Cairncross, S. "Effect of Washing Hands with Soap on Diarrhoea Risk in the Community: A Systematic Review," *Lancet Infectious Diseases*. 2003, *3*(5), 275–281.

De Silva, P.; Rachman, S. *Obsessive-Compulsive Disorder,* 2nd ed.; Oxford University Press: Oxford, 1998.

Didier Pittet, MD, MS; Anne Simon, MD; Stéphane Hugonnet, MC, MSc; Carmen Lúcia Pessoa-Silva, MD; Valérie Sauvan, RN; and Thomas V. Perneger, MC PhD. "Hand Hygiene among Physicians: Performance, Beliefs, and Perceptions," *Annals of Internal Medicine*. 6 July 2004, *141*(1), 1–9.

Doctor's Guide Publishing Limited Website. Antibacterial Products May Worsen Problem of Resistant Bacteria. www.docguide.com/news/content.nsf/news/ (accessed April 24, 2006).

Earthlife Web Website. Bacteria and Disease. www.earthlife.net/prokaryotes/disease.html (accessed April 26, 2006).

Encyclopædia Britannica Website. Lister, Joseph, Baron Lister, of Lyme Regis. www.britannica.com (accessed April 24, 2006).

Encyclopædia Britannica Website. Semmelweis, Ignaz Philipp. www.britannica.com (accessed April 24, 2006).

Encyclopædia Britannica Website. Soap and Detergent. www.britannica.com/eb/article?eu=115215 (accessed April 24, 2006).

EngenderHealth Website. Online Courses. Infection Prevention. Handwashing. www.engenderhealth.org (accessed April 24, 2006).

Fox News Channel Website. Dirt-Asthma Link Needs Scrubbing. www.foxnews.com (accessed May 3, 2006).

The Global Public-Private Partnership for Handwashing with Soap Website. Handwashing Can Save Millions of Lives. www.globalhandwashing.org (accessed April 24, 2006).

HowStuffWorks Website. Bacteria. health.howstuffworks.com/define-bacteria.htm (accessed April 26, 2006).

Institute for the Study of Healthcare Organizations & Transactions Website. Health & Behavior. Hot Topics. Hand Washing by Health Care Providers. www.institute-shot.com (accessed July 14, 2006).

Kimberly-Clark Australia Website. Research. Hand Hygiene. www.keepfoodsafe.com/research.html (accessed April 24, 2006).

The Medical Journal of Australia Website. The Contagiousness of Childbed Fever: A Short History of Puerperal Sepsis and Its Treatment. www.mja.com.au/public/issues/177_11_021202/dec10354_fm.html (accessed April 26, 2006).

MyDr.com Website. Asthma and the Hygiene Hypothesis. www.mydr.com.au (accessed May 3, 2006; need to create login name to access article).

The Nemours Foundation Website. For Kids. Dealing with Feelings. Obsessive-Compulsive Disorder. kidshealth.org (accessed July 14, 2006).

The Nemours Foundation Website. For Parents. First Aid & Safety. Home Safety & First Aid. Why Is Handwashing So Important? kidshealth.org (accessed May 11, 2006).

The Nemours Foundation Website. For Kids. Kids Talk. Q & A. What Are Germs? kidshealth.org (accessed May 11, 2006).

NurseWeek Publishing, Inc. Website. Wash Out: Could Antibacterial Soaps Create New Bacterial Strains? www.nurseweek.com/features/98-10/soap.html (accessed April 24, 2006).

O'Neil, J. Vital Signs: Got Germs? Regular Soap Will Do. *The New York Times,* May 12, 2006.

PreventDisease Website. Common Household Antibacterial Found Ineffective and Harmful. preventdisease.com/news/articles/antibacterial_found_ineffective.shtml (accessed April 24, 2006).

PreventDisease Website. 'Wash Your Hands' Signs Only Work for Women: Study. preventdisease.com (accessed April 24, 2006).

Respiratory Research Website. The Coming-of-Age of the Hygiene Hypothesis. respiratory-research.com (accessed May 3, 2006).

Salon.com Website. Talking Dirty: Bring on the Germs. Too Much Cleanliness May Be Making Some People Sick. dir.salon.com/story/health/feature/2000/05/03/germ_warfare/index.xml (accessed May 3, 2006).

Scientific American Website. Navy Recruits Wash Their Hands of Coughs and Colds. www.sciam.com (accessed April 24, 2006).

Silicon Valley/San Jose Business Journal Website. Antibacterial Soaps Presenting New Kind of Hygiene Problem. sanjose.bizjournals.com/sanjose/stories/2001/04/02/focus2.html (accessed April 24, 2006).

The Soap and Detergent Association Website. Hand Hygiene. 2005 Hand Hygiene Survey. www.cleaning101.com/HandHygiene (accessed April 26, 2006).

The Soap and Detergent Association Website. SDA Kids Corner. The History and Chemistry of Soaps and Detergents. www.cleaning101.com (accessed May 12, 2006).

The State News Website. Bar Soap in Bathrooms Rubs Some the Wrong Way. www.statenews.com/editionsspring99/040899/p1_soap.html (accessed April 24, 2006).

Sweeney, D. "Berry Full of DNA," *Biology: Exploring Life;* Pearson Education, 2001.

U.S. Food and Drug Administration Website. The Battle of the Bugs: Fighting Antibiotic Resistance. www.fda.gov (accessed April 24, 2006).

U.S. Food and Drug Administration Website. Food. Women's Health. Pre-school. In Day-Care Centers, Cleanliness Is a Must. www.fda.gov (accessed May 12, 2006).

U.S. Food and Drug Administration Website. Soap. www.cfsan.fda.gov/~dms/cos-215.html (accessed April 24, 2006).

Virtual Museum of Bacteria Website. Pathogenic Bacteria. Pathogenicity. www.bacteriamuseum.org (accessed April 26, 2006).

Weiss, S.T. "Eat Dirt—The Hygiene Hypothesis and Allergic Diseases," *The New England Journal of Medicine.* 2002, *347*(12), 930–931.

Word Spy Website. Hygiene Hypothesis. www.wordspy.com (accessed May 3, 2006).

## Water Purification

Advanced Buildings Technology & Practices Website. Cisterns/Rainwater Harvesting Systems. advancedbuildings.org (accessed June 1, 2006).

American Water Works Association Website. A Brief History of Drinking Water. www.awwa.org (accessed June 1, 2006).

American Water Works Association Website. Fact Sheets: Cryptosporidium. www.awwa.org/advocacy/pressroom/crypto.cfm (accessed June 1, 2006).

American Water Works Association Website. 25 Facts About Water. www.awwa.org (accessed June 1, 2006).

American Water Works Association Website. Parts Per Million. www.awwa.org (accessed June 1, 2006).

Best Stuff.com Website. Background on Fitness Waters. www.beststuff.com/article.php3?story_id=3599 (accessed June 1, 2006).

Bottled Water Web Website. GULP! Bottled Water Is Number Two, and Enhancements Are Coming. www.bottledwaterweb.com (accessed June 25, 2004; requires membership).

Bottled Water Web Website. Reading Between the Lines of Bottled Water Labels. www.bottledwaterweb.com/articles/avw-0002.html (accessed July 29, 2003; requires membership).

Bottled Water Web Website. Regulations. www.bottledwaterweb.com/regulations.html (accessed July 29, 2003; requires membership).

Bottled Water Web Website. A Technical Analysis of the NRDC Report on Bottled Water. www.bottledwaterweb.com (accessed July 29, 2003; requires membership).

Bottled Water Web Website. Water Products Categorization Chart. www.bottledwaterweb.com/waterchart.html (accessed July 29, 2003; requires membership).

Brody, J. A Dietary Mineral You Need (and Probably Didn't Know It). The New York Times, May 18, 2004.

Brown, T.L.; LeMay, H.E., Jr.; Bursten, B.E. *Chemistry: The Central Science*; Prentice Hall: Englewood Cliffs, NJ, 1994.

Canadian International Development Agency Website. Clean Water. www.acdi-cida.gc.ca (accessed June 1, 2006).

Cancer Consultants.com Website. Managing Side Effects. Electrolyte Imbalance. patient.cancerconsultants.com (accessed June 2, 2006).

Centers for Disease Control and Prevention Website. Diseases & Conditions. Water-Related Diseases. Cryptosporidiosis. www.cdc.gov (accessed June 1, 2006).

Cocanour, B.; Bruce, A. "Osmosis and the Marvelous Membrane," *Journal of College Science Teaching*. November 1985, 127–130.

Cohn, D. Lead in D.C. Water Slashed. *The Washington Post*, May 21, 2004, B01.

Curry, J. What's the Deal with 'Enhanced Water'? *The News & Observer*, September 26, 2002.

Federal Citizen Information Center Website. Drinking Water From Household Wells. www.pueblo.gsa.gov (accessed June 1, 2006).

Fumento, M. Bottled Fear Peddled by the NRDC. *The Washington Times*, April 8, 1999.

Fumento, M. Dirty Water: Will the United States Repeat Peru's Chlorine Folly. *Reason Magazine*, May 1996.Heath R. *Basic Ground-Water Hydrology*; Water-Supply Paper 2220; U.S. Geological Survey: Washington, DC, 1989.

Hill, J.W.; Kolb, D.K. *Chemistry for Changing Times*, 11th ed.; Prentice Hall: New York, 2006.

History Link 101 Website. History of Farming. Story of Farming. www.historylink101.com (accessed June 1, 2006).

Holtzclaw, H.; Robinson, W.; Odom, J. *General Chemistry*; Heath: Lexington, MA, 1991; pp 393–394.

Howard Hughes Medical Institute Website. Ask a Scientist. Please Explain the Physiology of Thirst. What Mechanisms Does the Body Use to Produce the Sensation of Thirst. www.askascientist.org (accessed June 1, 2006).

HowStuffWorks Website. How Your Kidneys Work. www.howstuffworks.com (accessed May 26, 2006).

HowStuffWorks Website. What Are Electrolytes? www.howstuffworks.com (accessed June 1, 2006).

Innvista Website. Electrolytes. www.innvista.com (accessed June 1, 2006).

Institute of Medicine of the National Academies Website. Food & Nutrition. Dietary Reference Intakes: Water, Potassium, Sodium, Chloride, and Sulfate. www.iom.edu (accessed June 1, 2006).

International Bottled Water Association Website. Bottled Water Facts. FAQs. www.bottledwater.org (accessed May 26, 2006).

International Bottled Water Association. Statement of the International Bottled Water Association (IBWA) Related to Guidance to Drink Eight, 8-ounce Servings of Water Each Day. www.bottledwater.org/public/2002_Releases/8of8statement.htm (accessed June 1, 2006).

Iowa State University Extension Website. Bottled Water—To Drink or Not to Drink. www.extension.iastate.edu (accessed June 2, 2006).

Kansas State University Website. Nutrition Spotlight. Water—The Body's Most Valuable Liquid Asset. www.k-state.edu (accessed June 1, 2006).

Kolata, G. New Advice to Runners: Don't Drink the Water. *The New York Times*, May 6, 2003, 5.

Mahon, B.; Mintz, E.; Greene, K.; Wells, J.; Tauxe, R. "Cholera Epidemic in U.S. Courtesy of EPA 'Science'," *The Journal of the American Medical Association*, 1996, 276, 307–312.

Marieb, E. *Human Anatomy & Physiology*; Benjamin Cummings: San Fransisco, 2001.

Marini, R. Water, Water Everywhere: Is the 8-glasses Rule Outdated and All Wet? *San Antonio Express-News*, July 19, 2004, 1C.National Resources Defense Council Website. Bottled Water: Pure Drink or Pure Hype? www.nrdc.org (accessed May 26, 2006).

Medical News Today Website. Coca Cola Withdraws Dasani, Its Purified Water from U.K. Market. www.medicalnewstoday.com (accessed May 26, 2006).

MicrobeWorld Website. Drinking Water Treatment. www.microbeworld.org (accessed June 1, 2006).

Nestle Website. The History of Bottled Water. www.nestle.com (accessed June 1, 2006).

Noonan, E. Children's Teeth May Suffer from Bottled Water Boom. *Associated Press*, November 2, 1998.

OhNo! News Website. Perrier. www.ohnonews.com/perrier.html (accessed May 26, 2006).

Oliphant, J.A.; Ryan, M.C.; Chu, A. "Bacterial Water Quality in the Personal Water Bottles of Elementary Students," *Canadian Journal of Public Health*. September/October, 2002, 93(5), 366–367.

The Phoenix Website. Archived Boston Content. Hot Water: The Latest in Designer Drinks: Hip H2O. www.thephoenix.com (accessed May 26, 2006).

Poland Spring Website. About US Poland Spring. Learn Our History. www.polandspring.com (accessed June 1, 2006).

Rolls, B.; Rolls, E. *Thirst*; Cambridge University Press: London, 1982. Scientific American.com Website. Bottled Twaddle: Is Bottled Water Tapped Out? www.sciam.com (accessed May 26, 2006).

Shakashiri, B. *Chemical Demonstrations*; University of Wisconsin Press: Madison, WI; Vol. 3, pp 283–285.

Squires, S. Enhanced Waters Are All Wet. *The Washington Post*, July 28, 2002.

Stark, P. *Last Breath: Cautionary Tales From the Limits of Human Endurance*; Ballentine: New York, 2001.

The Straight Dope Website. Can Water Be Too Pure? Is Too Much Water Bad for You? www.straightdope.com (accessed June 1, 2006).

Terrific Science Website. Lesson & Lab Exchange; Middle School—Risks & Choices Academy 2: Environmental Toxins. Introduction to Groundwater Hydrology. www.terrificscience.org (accessed May 26, 2006).

U.S. Environmental Protection Agency. "25 Years of the Safe Drinking Water Act: History and Trends," Office of Water: EPA 816-R-99-007, December 1999.

U.S. Environmental Protection Agency Website. Ground Water & Drinking Water. Drinking Water Contaminants. www.epa.gov (accessed June 1, 2006).

U.S. Environmental Protection Agency Website. Ground Water & Drinking Water. What Contaminants May Be Found in Drinking Water. www.epa.gov (accessed June 1, 2006).

U.S. Environmental Protection Agency Website. Mid-Atlantic Region: Lead in Washington, DC Drinking Water. www.epa.gov/dclead/disinfection.htm (accessed June 1, 2006).

U.S. Environmental Protection Agency Website. Water on Tap: What You Need to Know. www.epa.gov (accessed June 1, 2006).

U.S. Food and Drug Administration. Bottled Water: Better Than the Tap? www.fda.gov (accessed May 26, 2006).

U.S. Food and Drug Administration Center for Food Safety and Applied Nutrition Website. Bottled Water Regulation and the FDA. www.cfsan.gov/~dms/botwatr.html (accessed May 26, 2006).

U.S. Geological Survey Website. Water Science for Schools. Domestic Water Use. ga.water.usgs.gov/edu/wups.html (accessed May 26, 2006).

U.S. Geological Survey Website. Water Science for Schools. Ground Water Use in the United States. ga.water.usgs.gov/edu/wups.html (accessed May 26, 2006).

U.S. Geological Survey Website. Water Science for Schools. Public-Supply Water Use. ga.water.usgs.gov/edu/wups.html (accessed May 26, 2006).

University of California Davis Website. Hydration & Exercise. www.ucdmc.ucdavis.edu (accessed June 1, 2006).

University of Colorado Geography Department Website. Medical Geography and Cholera in Peru. www.colorado.edu/geography/gcraft/warmup/cholera/cholera.html (accessed June 1, 2006).

Urban Legends Reference Pages Website. Eight Glasses. www.snopes.com/toxins/water.htm (accessed June 1, 2006).

USA Today Website. Enhanced Waters Pour Onto Shelves. www.usatoday.com/money/advertising/2002-08-22-water_x.htm (accessed June 1, 2006).

Valtin H. "'Drink at Least Eight Glasses of Water a Day.' Really? Is There Scientific Evidence for '8 x 8'?" *Journal of the American Physiological Society*. August 2002.

Verbin, D. "A Brief History of Bottled Water." *Excaliber*, April 6, 2005. www.excal.on.ca (accessed June 1, 2006).

The Water FAQ Website. Water Basics. www.softwater.com/faq.html (accessed May 26, 2006).

Water and Wastes Digest Website. Enhanced Bottled Water. www.wwdmag.com (accessed June 1, 2006).

WaterHistory.org Website. Water and Wastewater Systems in Imperial Rome. waterhistory.org (accessed June 1, 2006).

*World Book Encyclopedia*, Millennium ed. V1, 2000, 581.

*World Book Encyclopedia*, Millennium ed. V5, 2000, 156.

World Health Organization Website. Water Sanitation and Health. Water-Related Diseases. Diarrhoea. www.who.int (accessed June 1, 2006).

World Resources Institute Website. Box 1.6 Cholera Returns. www.wri.org (accessed June 1, 2006).

# Healthy Air

"A Second Look at the Asthma Epidemic," *The Washington Times*, Editorial, April 16, 1998.

Aerias Air Quality Sciences Website. Odor and Odor Thresholds. www.aerias.org/DesktopModules/ArticleDetail.aspx?articleId=56 (accessed June 20, 2006).

Air Filtration Systems Website. Purafil® Library. Technical Paper. IAQ/Odor. Carbon Dioxide Monitoring: IAQ Indicator or Just Another "Boy Who Cried Wolf"? www.afslasvegas.com (accessed June 21, 2006).

American College of Allergy, Asthma & Immunology. Editorial Background. Allergy. www.acaai.org (accessed June 20, 2006).

American Council on Science and Health Website. "Sick Building Syndrome": A Diagnosis in Search of a Disease. www.acsh.org (accessed June 21, 2006).

American Industrial Hygiene Association Website. Is Air Quality a Problem in My Home? www.aiha.org (accessed September 15, 2006).

Apte, M.G.; Fisk, W.J.; Daisey, J.M. "Indoor Carbon Dioxide Concentrations and SBS in Office Workers," Proceedings of Healthy Buildings 2000 Conference. August, 2000.

Aukland Allergy Clinic Website. Vasomotor Rhinitis (VMR) or Idiopathic Non-Allergic Rhinitis. www.allergyclinic.co.nz (accessed June 20, 2006).

Backyard Nature Website. Fungi. Bread Mold Fungus. www.backyardnature.net (accessed June 20, 2006).

BBC News Website. On This Day. 21 August. 1986: Hundreds Gassed in Cameroon Lake Disaster. news.bbc.co.uk/ (accessed June 21, 2006).

BusinessWeek Website. Is Your Office Making You Sick? June 5, 2000. www.businessweek.com (accessed June 20, 2006).

California Research Bureau Website. Molds, Toxic Molds, and Indoor Air Quality. www.library.ca.gov/html/statseg2a.cfm (accessed June 22, 2006).

Centers for Disease Control and Prevention. Mold. Questions and Answers on *Stachybotrys* and other Molds. www.cdc.gov (accessed June 22, 2006).

Centers for Disease Control and Prevention Website. National Center for Infectious Diseases. Legionnaire's Disease. www.cdc.gov/ncidod (accessed June 20, 2006).

Chemical and Engineering News Website. Science and Technology. February 16, 2004. Mold Risks. www.cen-online.org (accessed June 21, 2006).

Childhood Asthma Foundation Website. About Asthma. www.childasthma.com (accessed June 21, 2006).

Clemson Extention Home and Garden Information Center Website. Dust Mites. hgic.clemson.edu (accessed July 10, 2006).

CNN.com Website. Health. Asthma and the Cockroach Connection. www.cnn.com (accessed June 21, 2006).

CNN.com Website. Health. Mold in the Home: Health Hazard or Hype? www.cnn.com (accessed June 21, 2006).

Consumer Reports.org® Website. Appliances. The Dubious Claims of Air Cleaners. Air Cleaners: Some Do Little Cleaning. www.consumerreports.org (accessed June 20, 2006).

Duhme, H.; Weiland, S.; Keil, U. "Epidemiological Analyses of the Relationship between Environmental Pollution and Asthma," *Toxicology Letters*. December 28, 1998, 307–316.

Fisk, W.J. "Health and Productivity Gains from Better Indoor Environments and Their Implications for the U.S. Department of Energy," *Annual Review of Energy and the Environment*. 25, 2000, 537–566.

Fumento, M. "Scents and Senselessness," *The American Spectator*. April 2000.

Goodwin Procter Website. Publications. Environmental Law Advisory. Indoor Toxic Mold: A Mushrooming Problem? www.goodwinprocter.com (accessed June 22, 2006).

HowStuffWorks Website. How Radon Works. www.howstuffworks.com (accessed June 21, 2006).

JunkScience.com Website. A Second Look at the Asthma Epidemic. www.junkscience.com (accessed June 21, 2006).

KidsHealth Website. For Kids. My Body. Skin. www.kidshealth.org (accessed July 10, 2006).

Kim, C.S.; Lim, J.Y.; Hong, C.S.; Shin, D.C. "Effects of Indoor $CO_2$ Concentrations on Wheezing Attacks in Children," Proceedings: Indoor Air 2002; July 2002.

May, J.C. *My House Is Killing Me!*; Johns Hopkins University Press: Baltimore, 2001.

Money, N.P. *Carpet Monsters and Killer Spores*; Oxford University Press: New York, 2004.

MSNBC Website. Family Wins Record Settlement Over Toxic Mold. www.msnbc.com (accessed June 22, 2006).

National Safety Council Website. Biological Contaminants. www.nsc.org (accessed June 20, 2006).

National Safety Council Website. Environmental Tobacco Smoke. www.nsc.org (accessed June, 21, 2006).

National Safety Council Website. Indoor Air Quality in the Home. www.nsc.org/ehc/indoor/iaq.htm (accessed June 20, 2006).

National Safety Council Website. Resources. Fact Sheets. In the Environment. Sick Building Syndrome. www.nsc.org (accessed June 20, 2006).

The Ohio State University Extension Website. Fact Sheets. Community Development Series. Environmental Tobacco Smoke. ohioline.osu.edu (accessed June 20, 2006).

The Ohio State University Extension Website. Fact Sheets. Community Development Series. Multiple Chemical Sensitivity. ohioline.osu.edu (accessed June 20, 006).

The Ohio State University Fisher Center for Real Estate Education and Research. Toxic Mold: A Growing Problem for the Real Estate Industry. www.fisher.osu.edu (accessed June 21, 2006).

Orange Coast Inspection Website. Toxic Mold. Health Effects. What Are the Potential Health Effects From Mold and Other Biological Pollutants. www.ocinspection.com (accessed June 20, 2006).

Permanent Website. NASA CELSS Work. NASA BioHome. www.permanent.com/s-ce-nas.htm (accessed June 20, 2006).

Philip Morris USA Website. Smoking & Health Issues. Secondhand Smoke. www.philipmorrisusa.com (accessed June 20, 2006).

Postgraduate Medicine online Website. Carbon Monoxide Poisoning. www.postgradmed.com (accessed June 21, 2006).

Quackwatch Website. Multiple Chemical Sensitivity. www.quackwatch.org (accessed June 20, 2006).

RadonSeal Website. Moisture and Indoor Air Quality. www.radonseal.com/moisture-indoor-air.htm (accessed June 21, 2006).

Raven P.H.; Johnson, G.B. *Biology*; Times Mirror/Mosby College: St. Louis, 1989, 632.

Science Education Partners Website. Mold Facts and Experiments. www.seps.org/cvoracle/faq/mold.html (accessed June 20, 2006).

The Straight Dope® Website. Is "Toxic Mold" the Next Environmental Threat? www.straightdope.com (accessed June 22, 2006).

Stutte, G.W. and Wheeler, R.M. "Accumulation and Effect of Volatile Organic Compounds in Closed Life Support Systems," *Advances in Space Research*. 20(10), 1997, 1913–1922.

U.S. Environmental Protection Agency Website. An Introduction to Indoor Air Quality. www.epa.gov/iaq/ia-intro.html (accessed June 20, 2006).

U.S. Consumer Products Safety Commission Website. Indoor Air Pollution: Introduction for Health Professionals. Two Long-Term Risks. Asbestos and Radon. www.cpsc.gov (accessed June 21, 2001).

U.S. Environmental Protection Agency Website. A Brief Guide to Mold, Moisture, and Your Home. www.epa.gov (accessed June 22, 2006).

U.S. Environmental Protection Agency Website. Asthma and Indoor Environments. Indoor Environmental Asthma Triggers. Dust Mites. www.epa.gov (accessed July 10, 2006).

U.S. Environmental Protection Agency Website. Indoor Air Quality (IAQ). www.epa.gov/iaq/ (accessed June 20, 2006).

U.S. Environmental Protection Agency Website. Publications. Indoor Air Quality. Consumer's Guide to Radon Reduction. www.epa.gov (accessed July 12, 2006).

U.S. Environmental Protection Agency Website. Publications. Indoor Air Quality. Protect Your Family and Yourself from Carbon Monoxide Poisoning. www.epa.gov (accessed June 21, 2006).

U.S. Environmental Protection Agency Website. Indoor Environment Management. Children's Health Initiative: Toxic Mold. www.epa.gov (accessed June 22, 2006).

U.S. Environmental Protection Agency Website. Project A.I.R.E. Reading Material: Health Effects. www.epa.gov/region01/students/teacher/aire.html (accessed June 20, 2006).

U.S. Environmental Protection Agency Website. Project A.I.R.E. Reading Material: Indoor Air Quality. www.epa.gov/region01/students/teacher/aire.html (accessed June 20, 2006).

U.S. Environmental Protection Agency Website. IAQ Design Tools for Schools. www.epa.gov/iaq/ (accessed June 21, 2006).

U.S. Environmental Protection Agency Website. Publications. Indoor Air Quality. Indoor Air Pollution: An Introduction for Health Professionals. www.epa.gov (accessed June 20, 2006).

U.S. Environmental Protection Agency Website. Publications. Indoor Air Quality. The Inside Story: A Guide to Indoor Air Quality. www.epa.gov (accessed June 20, 2006).

U.S. Environmental Protection Agency Website. Publications. Indoor Air Quality. Ozone Generators That Are Sold as Air Cleaners: An Assessment of Effectiveness and Health Consequences. www.epa.gov (accessed June 20, 2006).

U.S. Environmental Protection Agency Website. Publications. Indoor Air Quality. Targeting Indoor Air Pollution: EPA's Approach and Progress. www.epa.gov (accessed June 20, 2006).

University of Kentucky Extension Service Website. College of Agriculture. Indoor Air Pollutants: Detection and Control Measures. www.ca.uky.edu (accessed June 21, 2006).

*What's That Smell? The Science Behind Adolescent Odors*, Mickey Sarquis, Ed.; Terrific Science Press: Middletown, OH, 2003.

Wolverton Environmental Services, Inc. Website. Indoor Air Pollution. www.wolvertonenvironmental.com (accessed June 21, 2006).

## Healthy Skin

ABC News Website. Tanners Seek UV Rays Despite Danger. abcnews.go.com (accessed May 2, 2006).

Altruis Biomedical Network Website. Skin Lighteners. www.pigmentation.net/bleach.html (accessed April 28, 2006).

American Academy of Dermatology Website. 2006 Melanoma Fact Sheet. www.aad.org (accessed April 29, 2006).

American Academy of Dermatology Website. 2006 Skin Cancer Fact Sheet. www.aad.org (accessed April 29, 2006).

American Cancer Society Website. Do We Know What Causes Nonmelanoma Skin Cancer? www.cancer.org (accessed April 28, 2006).

American Cancer Society Website. Tanning Lamps Increase Skin Cancer Risk. www.cancer.org (accessed April 28, 2006).

American Cancer Society Website. What About Tanning Pills and Other Tanning Products. www.cancer.org (accessed April 28, 2006).

American Cancer Society Website. What Are the Risk Factors for Nonmelanoma Skin Cancer? www.cancer.org (accessed April 28, 2006).

American Academy of Family Physicians Website. Early Detection and Treatment of Skin Cancer. www.aafp.org (accessed April 29, 2006).

Archives of Dermatology Website. Youth Access Laws: In the Dark at the Tanning Parlor. archderm.ama-assn.org (accessed May 2, 2006).

BBC News Website. Senegal Doctors Demand Skin Cream Ban. news.bbc.co.uk (accessed May 2, 2006).

Bluhm, R.; Branch, R.; Johnston, P.; Stein, R. "Aplastic Anemia Associated with Canthaxanthin Ingested for 'Tanning' Purposes," *Journal of the American Medical Association*. 1990, *264*, 1141–1142.

Calcutta, India, Telegraph Website. Ads Cleansed in All Fairness: Commercials Pulled off Television After Viewer Objections. www.telegraphindia.com (accessed May 2, 2006).

California Newsreel Website. Race—The Power of Illusion: Race Literacy Quiz. www.newsreel.org (accessed April 26, 2006).

California Newsreel Website. Race—The Power of Illusion: Ten Things Everyone Should Know About Race. www.newsreel.org (accessed April 28, 2006).

California Newsreel Website. Race—The Power of an Illusion:Discussion Guide Toolkit. www.newsreel.org (accessed April 28, 2006).

Centers for Disease Control and Prevention: National Center for Health Statistics Website. Faststats A to Z. Deaths/Mortality. Deaths—Leading Causes. www.cdc.gov/nchs (accessed April 29, 2006).

Centers for Disease Control and Prevention Website. National Vital Statistics Reports. www.cdc.gov (accessed April 28, 2006).

Chen, I. "The Biology of Sunscreen: New Ways to Stop the Rays," *Discover*. 2003, *24*(6).

The Christian Science Monitor Website. Archive. Africans Look for Beauty in Western Mirror: Black Women Turn to Risky Bleaching Creams and Cosmetic Surgery. www.csmonitor.com (accessed April 28, 2006).

CNN Website. Health. Tan Now, Pay Later. www.cnn.com (accessed April 28, 2006).

CNN Website. Skin Deep: Dying To Be White. www.cnn.com (accessed May 2, 2006).

The Darker Side of Tanning. www.fda.gov (accessed May 1, 2006).

Dellavalle, R.P.; Parker, E.R.; Cersonsky, N.; Hester, E.J.; Hemme, B.; Burkhardt, D.L.; Chen, A.K.; Schilling, L.M. "Youth Access Laws: In the Dark at the Tanning Parlor?" *Archives of Dermatology*. 2003, *139*, 443–448.

Doctor's Guide Website. Bleaching Products Influence Skin Disease Presentation. www.docguide.com (accessed May 2, 2006).

EMedicine Website. Skin Lightening/Depigmenting Agents. www.emedicine.com (accessed May 2, 2006).

Federal Trade Commission Website. Indoor Tanning. www.ftc.gov (accessed April 26, 2006).

Fitzpatrick, T.B. "The Validity and Practicality of Sun-Reactive Skin Types I through VI," *Archives of Dermatology*. 1988, *124*, 869–871.

Gazette.Net Website. Race: Anthropologists Say Divisions Were Made by Man. www.gazette.net (accessed April 28, 2006).

Godiva Skin Care Website. Product Info. Whitening. Godiva Licorice Skin Whitening Products. www.godivaskincare.com/prodinfo.html (accessed May 2, 2006).

Hall, H.I.; Rogers, J. "Sun Protection Behaviors Among African Americans," *Ethnicity & Diseases*. 1999, *9*, 126–131.

HowStuffWorks Website. How Sunburns and Sun Tans Work. www.howstuffworks.com (accessed May 1, 2006).

HowStuffWorks Website. How Do Sunless-Tanning Products Work? www.howstuffworks.com (accessed April 26, 2006).

Internet Dermatology Website. Sun Damage and Prevention. www.telemedicine.org (accessed May 1, 2006).

Jamaicans.com Website. The Skin Bleaching Phenomenon in Jamaica. www.jamaicans.com (accessed April 28, 2006).

Journal of Chemical Education Website. Putting UV-Sensitive Beads to the Test. jchemed.chem.wisc.edu (accessed April 28, 2006).

Knight, J.M.; Kirincich, A.N.; Farmer, E.R.; Hood, A.F. "Awareness of the Risks of Tanning Lamps Does Not Influence Behavior Among College Students," *Archives of Dermatology*. 2002, *138*, 1311–1315.

Kovaleski, S.F. "In Jamaica, Shades of an Identity Crisis," *The Washington Post*. Aug 5, 1999, A15.

Lauden, L. *The Book of Risks*; Wiley: New York, 1994.

Looking Fit Website. Sunless Explosion. www.lookingfit.com (accessed May 1, 2006).

Marcus Garvey Library and Heritage Group Website. Education. Bleaching Creams and Chemical Peels. www.marcusgarveylibrary.org.uk (accessed April 28, 2006).

Media Awareness Network Website. The Price of Happiness: Student Questionnaire. www.media-awareness.ca (accessed April 28, 2006).

Michigan State University Website. Sunscreens and Sunblocks. www.msu.edu (accessed May 2, 2006).

Mid Day Multimedia, Ltd. Website. Fairness Cream Ad Is Vulgar, Discriminatory. web.mid-day.com (accessed April 28, 2006).

National Geographic Website. National Geographic News Photo: Skin Bleaching. www.nationalgeographic.com (accessed May 2, 2006).

National Geographic Website. Unmasking Skin. magma.nationalgeographic.com (accessed April 28, 2006).

National Weather Service Climate Prediction Center Website. Stratosphere. UV Index: Radiation Information and Climatological Statistics. UV Radiation and UV Index Information. www.cpc.ncep.noaa.gov (accessed April 26, 2006).

National Weather Service Climate Prediction Center Website. Stratosphere. UV Index: Radiation Information and Climatological Statistics. UV Radiation and UV Index Information. What Is the UV Index? Figure 2. www.cpc.ncep.noaa.gov (accessed April 26, 2006).

PBS Website. Race—The Power of an Illusion. What Is Race? www.pbs.org (accessed April 28, 2006).

PBS Website. Sorting People: Can You Tell Someone's Race by Looking at Them? www.pbs.org (accessed April 28, 2006).

People Living With Cancer Website. Melanoma. Risk Factors and Prevention. www.plwc.org (accessed April 29, 2006).

Roueche, B. "Annals of Medicine: A Good, Safe Tan;" *The New Yorker*. March 11, 1991. Samantha Brown Website. Health and Beauty. Are Skin Whiteners Safe? www.fourelephants.com (accessed May 2, 2006).

Science*Daily* Website. Smoking Triples Risk of a Common Type of Skin Cancer. www.sciencedaily.com (accessed April 29, 2006).

Serway, R.A. *Principles of Physics*; Harcourt Brace Jovanovich: Fort Worth, TX, 1994.

Sikes, R. "The History of Suntanning," *Journal of Aesthetic Science*. 1998, *1*(20).

Skin Cancer Foundation Website. African Americans Who Develop Melanoma Have a Worse Prognosis Than Whites. www.skincancer.org/news/010817-african.html (accessed April 27, 2006).

Skin Cancer Foundation Website. Sunscreen Limit of SPF 30 Is Questioned. www.skincancer.org/news/020625-pathak.php (accessed July 27, 2006).

South Asian Women's NETwork Website. Brown Skin and Sunscreen. www.sawnet.org (accessed April 28, 2006).

Stanford Solar Center Website. As the Sun Burns. solar-center.stanford.edu/webcast/wc03.html (accessed May 25, 2006).

Stanford Solar Center Website. For Students. What Is Ultraviolet Light? solar-center.stanford.edu (accessed April 28, 2006).

Statistics—Top 10 Causes Website. Causes of Death Younger Teens. www.statisticstop10.com (accessed May 24, 2006).

Sunless.com Website. Sunless and Safe. Sunless Tanning. Facts and Myths. www.sunless.com (accessed April 26, 2006).

Sunless.com Website. Sunless and Safe. Why Tanning Pills Don't Work. www.sunless.com (accessed April 26, 2006).

Suntanning.com Website. Tanning and Health. Tanning IS Healthy! www.suntanning.com (accessed April 28, 2006).

TanningTruth.com Website. The Fundamental Truth About Tanning. www.tanningtruth.com (accessed April 28, 2006).

TanningTruth.com Website. Positive Effects of UV Light. www.tanningtruth.com (accessed April 27, 2006).

ThingsAsian Website. White Is Beautiful in Vietnam. www.thingsasian.com (accessed April 28, 2006).

U.S. Environmental Protection Agency Website. Sunscreen: The Burning Facts. www.epa.gov/sunwise (accessed May 1, 2006).

U.S. Environmental Protection Agency Website. Sunwise School Program: UV Index. www.epa.gov (accessed April 28, 2006).

U.S. Food and Drug Administration Office of Cosmetics and Colors Website. DHA-Spray Sunless "Tanning" Booths. www.cfsan.fda.gov (accessed April 28, 2006).

U.S. Food and Drug Administration Office of Cosmetics and Colors Website. Sunscreens, Tanning Products, and Sun Safety. vm.cfsan.fda.gov (accessed April 28, 2006).

U.S. Food and Drug Administration Office of Cosmetics and Colors Website. Tanning Pills. www.cfsan.fda.gov (accessed April 28, 2006).

U.S. Food and Drug Administration Website. On the Teen Scene: Dodging the Rays. www.fda.gov (accessed April 28, 2006).

U.S. Food and Drug Administration Website. Sunscreens, Tanning Products, and Sun Safety. www.fda.gov (accessed May 2, 2006).

University of Colorado Health Sciences Center Website. Lack of Laws Limiting Use of Tanning Parlors Puts Teens at Risk. www.uchsc.edu (accessed May 2, 2006).

The Village Voice Website. Fade to White: Skin Bleaching and the Rejection of Blackness. www.villagevoice.com/news/0204,chisholm,31703,1.html (accessed May 1, 2006).

Women's eNews Website. Indian Women Criticize 'Fair and Lovely' Ideal. www.womensenews.org/article.cfm/dyn/aid/1308/context/archive (accessed April 26, 2006).